The Professional Skills Handbook for Engineers and Technical Professionals

The Professional Skills Handbook for Engineers and Technical Professionals

Kevin Retz

CRC Press
Taylor & Francis Group
Boca Raton London New York

CRC Press is an imprint of the
Taylor & Francis Group, an **informa** business

CRC Press
Taylor & Francis Group
6000 Broken Sound Parkway NW, Suite 300
Boca Raton, FL 33487-2742

Printed on acid-free paper

International Standard Book Number-13: 978-0-3674-2500-5 (Hardback)
978-0-3674-2499-2 (Paperback)

Visit the Taylor & Francis Web site at
http://www.taylorandfrancis.com

and the CRC Press Web site at
http://www.crcpress.com

Contents

Preface...ix
Acknowledgments and Thanks..xi
Author xiii
Introduction.. xv

Chapter 1   Lessons Learned and Food for Thought 1

        Learn to Say "I don't know" 1
        Be Careful What You Commit to "It's easier to get up a tree
        than down a tree" 2
        Newton's Third Law, "For every action, there is an opposite and
        equal reaction"...2
        Look for What's Missing in the Conversation or Not Being
        Recognized..3
        Always Give Your Best, Tenacity, and Persistence Matter 3
        Praise in Public and Criticize in Private 4
        Speak Up and Contribute 4
        Meeting Minutes Are a Must, So Are Action Logs 5
        Don't Put Lipstick on a Pig or Polish a Rock 5
        Clear Communication 6
        The Boss...6
        Integrity 7
        Represent Yourself, Represent Your Organization, Represent
        Your Company..8
        Respect 9
        Stress 9
        Reflection... 10
        Feedback.. 11
        Difficult Conversations.. 14
        Learning 16
        Questions 17
        Purpose/Passion 18

Chapter 2   Leading and Working with Your Team... 21

        Trust and Mutual Accountability 21
        A Common Shared Purpose.. 21
        Goals 22
        Skills.. 22
        Total Team...23

Chapter 3   Virtual Teams, Leading and Working with Virtual Teams 29

People and the Virtual Skill Set 29
Guidelines for Technology Use 31
Phone Calls and Text Messages 33
Email 33
Initial Team Meeting 34

Chapter 4   Developing and Delivering Presentations 37

Giving the Presentation 40
Relax 40
Multimedia 41

Chapter 5   Understanding Creativity and Innovation 43

Factors That Affect Creativity and Innovation within an
Organization 43
    Communication 46
    Training 47
    Mentoring 47
    Leadership 47
    Experimentation 48
    Strategy 48
    Challenging Work 49
    Technology Awareness 49
    Recognition 49
        Enabling a Creative Mindset 50
        Diverse Interests 50
        Background Noise 50
Seek Out Advice or Other Sources of Information 50
Exercise and Health 51
Build Creative Bridges 51
Build Expertise 52
Keep a Notepad 52
Characteristics of Creative Individuals 52

Chapter 6   Critical Thinking Skills 53

Critical Thinkers 53
Ways to Improve Critical Thinking Skills 58
    Lifelong Learning 58
    Develop a Questioning Mind 58
    Develop Active Listening Skills 58
Everyday Skills 60

Chapter 7 Developing Requirements, Goals, and Objectives 63

    Key Points 65
    Validation 66
    Tips for Managing a Requirements Document 67

Chapter 8 Developing and Understanding Strategy 69

    Strategic Thinking Skills/Tactics 71

Chapter 9 Root Cause Analysis/What to Do When Things Do Not Go as
Planned 75

Chapter 10 Lean Engineering the Basics 79

    Three Basic Principles of Lean 79
    Concepts or Practical Application of Lean Principles 80
        Muda 80
    Application of Lean Principles 82
        Value Stream Mapping 82
    Lean in Practice/The Lean Tool Box 83
        5S 83
        Andon 84
        Gemba 84
        Kaizen 85
        Poka-Yoke 85
        Standard Work 86
        Total Productive Maintenance 87
    Lean in Summary 87

Chapter 11 Cost Estimation/How Do You Figure Out the Cost of
Manufacturing 89

Closing Thoughts 95

Index 97

Preface

Today's engineers are technically sound upon graduation, but from the first day on the job, they are expected to be more than just an extension of a technical team. Today's engineer is expected to lead teams, provide support for cost savings measures, assist in marketing programs, help layout strategy or project requirements, and provide technical support to design and manufacturing teams. In today's environment, engineers need more than just technical expertise to not only survive, but thrive. This handbook covers those secondary skills that all good engineers need to master. It also contains a section on lessons learned. The lessons learned covers topics that all engineers will face in their careers and provides guidance to help navigate the topic being covered so the individual can find their own solution. In today's fast-paced technical environment, engineers need to combine technical expertise along with a wide range of secondary skills to survive. This handbook covers those secondary skills that are so valuable for an engineer's success.

Acknowledgments and Thanks

I wish to express my appreciation to my wife Marlo for her encouragement and support during the writing of this handbook, to my daughter Kelsey for her support and willingness to ask questions, and to Chris Holtorf for his support, encouragement, and professional/real-world technical insight which helped refine the final product. Without their support, this project would have taken longer and would not have been the product that it is. I also wish to thank Johnathan Plan, Bhavna Saxena, and Joel Stein for their guidance and encouragement during this project. I must also say thanks to Monica Felomina from Lumina Datamatics and her team for helping me make the final product shine.

Author

Kevin Retz, PhD, is an executive business leader, consultant, and engineering leader with over 25 years experience leading and guiding multi-million-dollar prototype aircraft design and assembly projects, 100+ team member composite shops, and executive coaching in competitive analysis, technology roadmapping, and technology integration. Dr Retz has a broad background in organizational development, organizational transformation, operations, and manufacturing. His specialty is creating a working environment centered on integrity, communication, and accountability.

Introduction

In today's engineering and manufacturing environment, engineers are more than just technical resources. Today's engineer is expected to lead teams, provide support for cost-saving measures, assist in marketing campaigns, and provide technical support to design and manufacturing teams. This handbook is intended to be a resource for today's engineer, technology leader, and scientist. It covers many topics that today's manufacturing engineer will face. Each topic is covered in a concise and straight-forward manner.

The chapters are brief and are not intended to turn the reader into an expert in any one area. This handbook is intended to be used as a reference guide and act as a starting point to expand knowledge and understanding in a particular area of interest or need. The chapters are kept short so that they are easy to read and understand. By keeping the chapters short, the author hopes to inspire the quest for a deeper under-standing of the subject.

In today's environment, engineers need more than just technical expertise to not only survive, but to thrive. This handbook covers those secondary skills that all good engineers need to master. It is based on nearly 30 years of experience of work-ing in the field of engineering and composites manufacturing, from starting out as a lay-up technician, to becoming an engineering manager and a composite shop man-ager. I have always been taught to give back and to share my knowledge and lessons learned. In a small way, I hope this handbook helps others and meets its intention of being a handy reference book.

I use the word engineer, but managers, cell leads, scientists, technicians, and stu-dents will all find this handbook useful. It is meant as a reference guide for any individual working in the field of engineering or management today, even if they are not working directly on the shop floor.

I hope this handbook meets its intent, that of being a handbook for anybody work-ing in the fields of engineering and technical services. Technical expertise is impor-tant, but in today's environment, the non-technical aspects of the job will set the good engineer apart from the great engineer. The great engineer has the ability to work in cross-functional teams and directly with shop personnel. The great engineer can run cross-functional, virtual teams and add value across the organization.

To achieve success in today's environment, engineers needs to be able to combine technical expertise with teamwork and other secondary skills, such as Lean and root cause analysis.

Thank you and good luck in all of your endeavors.

1 Lessons Learned and Food for Thought

In this chapter, I will talk about some lessons that I have learned over the years. These rules and lessons are usually not discussed in classes, but when they come up, and they will, you will wish they had been. Some of these are life lessons, and some of these are common sense that just need to be re-mentioned from time to time. Some of these I have learned and re-learned the hard way. Please learn from some of my mistakes or at least pause and think.

LEARN TO SAY "I DON'T KNOW"

Follow "I don't know" up with "but I can find the answer" or "let me find that answer out for you." We have all been in meetings when somebody has gone off on a long rant, it is obvious that they do not know a thing about the subject at hand. This rant makes them look foolish and, in many cases, lose respect and wastes time. We have all given presentations or have been in meetings and asked questions for which we don't know the answer. Don't try to bluff your way out of it. It will show to more than one individual in the room that you do not know the answer.

Instead of bluffing your way through the question and answer, "I don't know, but I can find the answer for you and report back to you at the next review," or in some other appropriate manner or time.

By answering in this fashion, you are giving yourself time to learn something new and you're not making a fool of yourself. If you try to gloss over or bluff your way through an answer, you will lose respect, validity in your knowledge, and your presentation will not be taken seriously. The audience will lose interest (or worse become agitated).

I will also caution about the tendency some people have of "winging it." Trying to make it look like they have the answer when they really don't. I have seen more than once how engineers try to inject information that they have heard from a second or third source as fact and firsthand knowledge. I have also seen engineers rattle off numbers, equations, and random facts to try to inject themselves into the process. In all cases, the engineer came off as a nuisance. Even Teflon wears out after time, and when something sticks to it, it's hard to get off. Think of that old Teflon-coated baking pan that you had to throw away after too many uses. You may get by winging it once, but it will catch up to you. Always stick to the facts, and when you don't know, say so and then offer to get the answer. If possible, defer the question to somebody that is an expert in the field. Let the expert take the question so that everybody can learn at the same time.

None of us have all the answers. We are only human, admit it. Take the action item, respond to the action item, and move on.

BE CAREFUL WHAT YOU COMMIT TO "IT'S EASIER TO GET UP A TREE THAN DOWN A TREE"

The old saying "It's easier to get up a tree than down a tree" I have heard in many similar forms. I don't know who has credit for it, but it is not me—"You can always fall or get knocked out of a tree. When you fall, it's going to hurt when you hit the ground or branches on the way down."

Think about the pictures of cats or bear cubs stuck in trees. Or if you ever climbed a tall tree as a little kid and thought once you got way up into it, how am I going to get down? Now, what this saying is telling us is be very careful what we commit to. Making commitments and taking on work assignments is great and required for job and professional growth.

But only commit when you have a clear understanding of what you're committing to. There are questions you need to ask. "What are the expectations?" "What is the deliverable?" "Do I have the time and the resources to follow through in the expected time frame?" When you commit to do something, always expect to be held accountable for that commitment. A commitment is your reputation.

When you make a schedule (which is the same as a commitment), think of all the players that should have input. When you're working on a design, think about the tooling team and the manufacturing team. Get their input to make sure you can tool and build what you're designing. You're going to look foolish when during the design review your operations lead states "I can't build that," usually followed by some very colorful forms of the English language. By involving all elements of the team early, you know you can make the schedule and hit your commitment.

When you make a commitment, you need to keep in mind that things will happen. Try to figure out what those can be before you make the commitment. If it's just finding an answer to a question, that is one thing. But if you need to involve others, keep that in mind and get them involved early. The other members on your team have their own commitments and their own bosses to please. It is not a good day when you go and talk to your boss and tell them that you could not complete a task because of things out of your control. "Results always speak better than excuses."

Remember, only make commitments when you have the bandwidth and the resources to hit that commitment. Commitments come in many forms: budget, schedule, memo, drawings, and document submittals.

Stick with your commitment. If problems come up, face these problems. If necessary, adjust the deliverable with the customer. Never just let a commitment slide. When commitments are not met or are constantly adjusted, your reputation will suffer.

NEWTON'S THIRD LAW, "FOR EVERY ACTION, THERE IS AN OPPOSITE AND EQUAL REACTION"

Newton's third law is true in life as well as physics. All actions have consequences, some good, and some may not be so good! Think through your actions and plan accordingly. This also goes with schedules and program plans. Use this same thought process as you set up a schedule and move through a program. Put milestones and checkpoints into the schedule so you can make sure you're making progress. Track your progress and, if needed, adjust in a timely manner.

Remember for every action, there is a reaction, you do one thing, something else cannot get done. If you take this course over this other course of action, the potential outcomes are different. If you select this requirement, then these systems will not meet specifications. If you chose this manufacturing process, then only these materials can be used.

I wonder if Mr. Newton knew his law would impact all aspects of life, not just physics, science, and engineering?

LOOK FOR WHAT'S MISSING IN THE CONVERSATION OR NOT BEING RECOGNIZED

People can always look at what is broken right in front of them and find a solution. It's finding the hidden broken or missing piece that is difficult. Humans like to focus on what is in front of them and what is apparent. Always think about what is not being shown and what is not visible. This is especially critical in design reviews, root cause analysis, and contract negotiations. By doing this during contract negotiations, it can help you drive to a better contract. During a program review or root cause analysis, it can help increase the solution space. You can also use this when touring a plant, look at your surroundings and what is not being discussed or pointed out. Look for the red tags on parts, the number of tools in rework, or the amount of inventory in work or sitting in job racks. You're not doing this to put anybody on the spot, use it to give clarity to the situation and to ask questions.

This approach can also be used when you are reviewing your own presentations and program plans. Look for the pros and the cons, all potential solutions, and actions. Make a list. Once the list is made, you can find answers to the questions and develop mitigation plans as needed.

ALWAYS GIVE YOUR BEST, TENACITY, AND PERSISTENCE MATTER

When given a task by your boss, just do it, even if you think it is beneath you. Set high standards for yourself and keep to those standards. Small tasks can be trials that lead to bigger and more visible assignments. When you make simple tasks shine, bigger assignments will follow.

When you are faced with challenges, don't give up, there is always a solution, so find it. Ask questions and don't be afraid to ask for help. The key is results, when you come up against a roadblock, get help and figure a way around it, through it, or over it, but do not just stop and quit. All projects will face challenges. When facing challenges, the key is to stay focused and work through the challenges with the same passion that you had at the start of the project. Don't let off the gas as you near the completion of a project. The hardest 10 yards are the 10 yards right before the goal line. Keep the pressure on and always complete the project to the best of your and your team's ability. If you show passion and drive, it will rub off on those around you.

Do not use the words "never" or "it's impossible" until you can show through hard work or analysis that the task cannot be done. Going to the moon was not possible nor was breaking the sound barrier, until a lot of hard work was put in and obstacles were overcome.

Always give your best, it will reflect well on you.

PRAISE IN PUBLIC AND CRITICIZE IN PRIVATE

In today's world, the statement "praise in public and criticize in private" is often forgotten. You will be at some presentation and somebody will just belittle the presenter. When this happens, think about how the presenter feels right then and what if that were you in their shoes. If a presenter is having a bad presentation, give some probing questions that will help them get on track. If you're a manager or a leader, give an excuse for a break to give the individual time to get over their nerves or talk to them in private and ask if they are truly ready for the presentation. Try to help the individual get out of the hole they may have dug.

Another scenario is when you have to talk to a team member that is not pulling their weight or an assignment that was just turned in and was not quality work. Your first reaction might be to send off an email to them and their boss and let them have it. Instead of the email to their boss, tell them upfront about the issue and give them a chance to correct the issue. If they cannot or will not correct the issue, then take it up with their manager, but remember to always stay above board and give the individual a chance to correct the issue. Conduct work always above board and fairly. Honesty and integrity will be noticed and remembered, and keep in mind, you may be working for that same individual in a few years.

When giving a critical comment or review always keep it about the situation and never about the individual. If discussing an area where the individual needs to improve give solid examples and never call the individual stupid or make them feel inferior. Always have critical conversations in private, never in an open forum where others may hear.

The flip side of this is when somebody does a good job or goes that extra mile let them know and let others know. Everybody likes pats on the back and peer recognition. When you let others know that you expect, recognize, and appreciate stellar work and performance, you will become the team leader or manager they want to work for or have on their team. When an individual does a great job, make sure that their boss and others on the team know. If possible, let their family know about the great job they did and the contribution it made, give them movie tickets or some small award. This extra step of giving recognition helps build not only confidence in the individual, but also respect.

SPEAK UP AND CONTRIBUTE

Don't be afraid, if you can add something to a conversation, speak up. Speaking up may be adding additional information or asking a question. Presentations are a good way to learn, so ask the question. Questions can trigger more discussion, which could lead to a better solution.

Caution: Do not ask questions or present additional information in a way that puts the presenter into a corner or makes them look bad. Don't be afraid to speak up and offer to take an action item. If questions are being asked or ideas are being kicked around and you can contribute to the conversation, make your voice heard and share your knowledge and ideas. Work is a participant sport, so get into the game. The question, idea, or thought that you state may be the spark that the team needs to come up with the great solution or breakthrough. If you can add to a conversation, then do it, and be confident in yourself. If you cannot add to the conversation, sit back, observe, and learn.

MEETING MINUTES ARE A MUST, SO ARE ACTION LOGS

I have learned this more than once: meeting minutes need to be done and sent out as soon as possible after all meetings. Meeting minutes are a pain, but serve some very important functions. Meeting minutes help people remember what was said and covered and what actions were taken and given out.

All meetings should serve several functions. First, inform the team what is going on and discuss next steps, help needed, and required actions. It is amazing how different people will remember different things or take away a different perspective from the same conversation. Meeting minutes helps everybody stay on the same page. If you see meeting minutes come out that you don't believe truly covered the meeting, take the time to get clarification before the team goes off in the wrong direction. Taking the time to get clarification and provide feedback saves time and frustration both for you and your teammates.

Action logs help people stay accountable. Sometimes, people work best on a deadline. Action logs that are clear, crisp, and have names and dates for each action can keep a team and the individuals on track and committed to the project. Action logs and meeting minutes are also a vehicle to give team and individual recognition for performance and accomplishments.

DON'T PUT LIPSTICK ON A PIG OR POLISH A ROCK

We have all or will work on a project that is just not going to meet the objectives, close the business case, or is no longer relevant, these projects are often times referred to as pigs. It may be a project that we have put a lot of work, sweat, and tears into! Let it go.

Managers and leaders need to recognize when a project is not going to go anywhere, and then talk to the team and either give them direction to make the project relevant or cut the project. As engineers, we need to recognize that a project we are working on is not meeting the objectives and either ask for help or ask to terminate the project. A pig will always be a pig even with a wig and lipstick. If the project is turning into a pig, we need to say so and stop the effort or get the assistance to get the project where it needs to go. If we are not sure, we need to ask for a review and see if others have the same conclusion.

Again, this is a saying I cannot take credit for, but understand it and live it. I have had to accept it myself or direct teams to stop working on a project when it was not going to give the desired results and future efforts could be utilized better elsewhere.

CLEAR COMMUNICATION

This is a hard one, communication written or spoken needs to be clear and concise. Make sure you say what you want to say in as few words as needed and make it clear. Boil the matter down to simple terms. There are several sayings over the years that cover this point, "keep it at an eighth-grade level," "know your elevator speech," "write your point on a 3 × 5 card." All of these are just telling you that time is critical in everybody's life, and you show respect when you don't waste time. People have a short attention span, so by keeping everything to the point you avoid the potential of confusion, misinterpretation, and giving the wrong impression. Confusion and misinterpretation will waste time, energy, and lead to frustration and heartburn.

Good communication is hard and takes time and practice. We all know people who seem to just love to hear themselves talk, the problem is these people are rarely if ever truly listened to. Strive to speak and write clearly. Good communication skills set leaders apart. This is one skill that can always be improved upon, so work at it, even in your email's, water cooler breaks, and presentations.

THE BOSS

Everybody, even the chief executive officer of a company, has a boss. Do not forget that you have a boss, and they have the right to be kept informed and take priority on your action list. Your manager/boss gives you assignments based on organizational needs. Keep in mind that your boss has a boss, and you do not know all the time where his assignments are coming from, those assignments may be coming from several pay grades up.

Take all assignments that your manager gives you and make those top priority, even if you do not want to do them. And always give them your best effort. If you need clarification on an assignment or you think your boss is heading in the wrong direction with an assignment then speak up about your questions or concerns, but at the end of the day, it's an assignment and work is work, so just buckle down and do it.

Another piece of advice when you're making decisions or working on a project, step back and take a company view or step up and take a perspective from your boss's viewpoint. This action will potentially change how you're looking at the project or your work. Stepping back and getting out of the trenches and the day to day grind, and taking a larger view of the landscape, can change how you view the assignment. If you need to make a course change, talk to your manager, they have an obligation to keep things moving and ensure that the groups objectives and goals are meet and your efforts are part of that bigger effort.

Do not take this concept of keeping the boss informed to mean that you have to allow them to micromanage, ask permission for every move you make, or to use it as a way to become the bosses pet. Your organization hired you to do a job based on your abilities, so do your work, but keep your boss informed as to how things are going, when you have questions or concerns, and when you need help. The last thing your manager wants is surprises. Do not let your boss find out that you ran into problems on a project during a review, or worse, from his boss. If this happens, it is not going to be a good day for you. Keep your boss informed and your life will be easier.

Managers were put into their current position for a reason and have experience. Leverage that experience when you can, and always, when you go to your boss with an issue, go with potential solutions and the pros and cons of those solutions. Your manager will appreciate that you brought up the issue, but will be impressed that you thought about it and brought potential options to the table to be discussed. Your options may not be the final solution, but they will show that you can think for yourself and be diligent in your work. Challenges and opportunities are just part of day-to-day living. Managers know this and want to know when something is not going as planned, and greatly appreciate it when options are brought to the table, as well as what the potential ramifications of the different actions are.

The military has a saying "show respect for the grade not necessarily the man." What this means is that you do not have to like or even respect the individual you're working for, but their position entitles them to a certain amount of respect and part of that is keeping them informed. Respect is earned, so do your job to keep your boss informed so they are never blindsided. If you find out you're working for a boss that you cannot respect, then it's time to look for a new manager and role.

INTEGRITY

Integrity should not even be discussed, it should be a given, but it is way too often forgotten by people up and down the organizational ladder. Everything you do should be above the board and done with integrity. Integrity is doing what needs to be done when nobody is watching to the best of your abilities and with respect for yourself, your organization, and the company.

Be fair and honest in all of your dealings, if you're not, it will come back and haunt you. Once you lose your integrity and the people around you feel like you cannot be trusted, then you're done. People will not want to work with you and managers won't want to give you any assignments that matter.

This goes for what you say, make sure that what you say in both verbal and written communication is true. Base statements of "fact" on facts and data that you can support and state assumptions and conjectures as such. If you are misstated or misquoted after the event, step in and clear up any misunderstanding or mis-interoperation. People are counting on you to speak clearly and factually and may act on those words or recommendations. Make sure your words, facts, and data are accurate.

Integrity will follow you and your name wherever you go. Keep your integrity and your name foremost in your thoughts and actions. You and only you own your name and the level of integrity associated with it, so make it count. If you are ever asked to do something unethical or suspect unethical behavior in your organization, then report it. If you feel that the organization at its core is unethical, then find another organization to work for, your name and reputation are too valuable. Some Enron employees carried the stigma of working for an "unethical organization" for a long time, even though they were not involved in or knew of the issues.

REPRESENT YOURSELF, REPRESENT YOUR ORGANIZATION, REPRESENT YOUR COMPANY

Your work and the work you're associated with have your name on it, so make sure it is a proper reflection of you. Always represent yourself in the best possible light. Make your name your brand, you have direct control over it, so make sure it shines by always putting your best effort forward.

Represent your organization, when you give a presentation, are on a tour, or interacting with other teams, remember your actions will reflect on you and your organization in some way. Team efforts and results reflect on its members, so give your teammates your best effort and expect it from them. If a team does not perform or is given a negative label, it reflects on every member of that team. The same is true if a team turns in a stellar performance, it will reflect well on each individual on the team and the team as a whole, the individual team members will become stars.

If you're a manager or a lead, take interest in the people within your organization as their level of performance will reflect on you. This should be a given in any leadership role. When your people have challenges, help them sort those challenges out. When individuals or teams have a major presentation to upper leadership, ask to see it before hand or have them do a dry run, especially if it is not an experienced individual team or if they don't like giving presentations. Help your peers and subordinates thrive and grow. When you coach and help others, you will become a person people will want to work for and with. You reflect your organization, and your organization reflects on you. Take pride in that organization.

Represent your company, you see people in society not take this one into consideration, which is sad. There are several aspects to this: first, the company has given you a job and, like you, that company has a name and a reputation, so when it comes to you, remember when you're in meetings with other companies (or organizations), in public, or at conferences that you do represent yourself and your company. When you're out with others, they will see you as a representative of your company. You may believe you're just speaking for yourself, but others will take your views and words as an indicator or a true representation of your company (or organization).

If you're in a true business setting be it in a meeting or a meal, do not make commitments for your company that you are not authorized to make. When you say you're going to do something, make sure it is within your work scope to make that commitment, because you are not only committing yourself, but your company and your collective reputations. Do not ever make commitments that are beyond your pay grade and ability to fully support.

I once went to an interview and was told, "You work for the Lazy X; your work ethic is poor." I asked this individual to look at my resume and my accomplishments. My accomplishments spoke for themselves. A company's reputation will follow you, so represent your company well, it will reflect on you. Her first impression of me was the same one that she had for all of the Lazy, guilt by association. I had to work to change her impression of me, and explain that her impression of the Lazy B should not be associated with me.

RESPECT

Respect is another one that should not need to be stated, but unfortunately does. Respect yourself and those around you, and treat everybody you meet with respect. When you show respect for others no matter who they are, it shows positively on your character, and people will want to work with you and for you. This does not mean to be a pushover. Expect the same treatment from others as well. We live in a global society, your actions and how you treat others will get around and be noticed, especially in the age of social media.

You may never know who you might meet, that person you just bumped into leaving the coffee shop may be the chief executive officer of the company you're pitching to in the afternoon. Or the individual you see straightening up the lobby may not be the receptionist, but the head of product development.

Another point of consideration, watch the manners of those around you, if somebody treats you with respect in meetings, but is rude to individuals from different cultures or to the waiter or flight attendant, then beware. That individual has a sense of superiority and entitlement. These are individuals you do not want to work for or with. And if you have the misfortune of having to work with them be careful, they will stab you in the back if it benefits them and not think twice.

Follow the golden rule and treat others as you would want to be treated, with respect and dignity.

STRESS

Today's world with its fast pace and constant demands is stressful. Stress can be triggered by any number of things, a presentation, trouble with a co-worker, or a family crisis. Here are a couple points of advice that I have used to help cope even though I have no magic bullet, stress is a fact of life and will affect different people in different ways.

When feeling stressed:

- Take a couple of deep breaths and clear your mind
- Exercise and eating right can help keep your mind focused and reduce stress
- Meditation has helped me at times and research shows that it helps. Sit quietly and try to clear your mind of any thought while breathing in slowly, hold the breath briefly, and then exhale. While meditating focus on the center of your forehead at the spot right between your eyebrows. Research has shown that meditating for five to twenty minutes a day can have some health benefits, as well as being a way to reduce stress
- Talk to a trusted friend, coach, colleague, or your significant other. Talking helps particularly if the person is a good listener. Discuss why you're feeling stressed or pressured. Often just the act of talking through a problem will help relieve the stress and allow you to think of and consider options
- Pets can help, research has shown that pets and the interaction with animals can lower stress. The caution here is to make sure you're ready for the responsibility of a pet. A pet can add stress to your life if you're not ready for the time commitment and responsibility of caring for one

- Step back and make a cup of tea and take a few moments to reflect and think before you act. The act of making a cup of coffee or tea and drinking it slowly forces you to take time out and think about the situation. This time helps you disengage from the issue and think about it in a clear manner and with a focused mind. Too often when we react out of frustration or anger we regret that action
- Reflection is a great way to reduce stress and see situations clearly
- Get over the image, be true to yourself, you know your passions and core values, stick to them. The caution here is that passions and purpose may change over time, but core values and beliefs stay constant for the most part. Core values give you that rock to hang onto when things get tough. When core values are strong, it helps you stay focused on what's important to you, thus reducing stress.
- Own who you are. Don't let others put labels on you unless you agree with that label. Own who you are and let it be an image you're comfortable with. This is a hard lesson at 23, a little easier at 55. The old saying "you need to love yourself first" is very true here
- Accept that you don't know everything. When you don't know something, admit it, and ask for help or go learn it, if you need to know the information or have an interest in it. Nobody knows everything, accept that in yourself and others
- Go for a walk and clear your mind. After you have cleared your mind, think about what it is that is causing you the stress. If you have a dog, take the dog for a walk, both of you will enjoy the exercise, and the walk will give you time to think and reflect.

Stress is part of life. How we handle stress is what matters. Stress can be a silent killer that can affect our health and relationships. I have no magic bullet to handle stress, each individual is different, the key is to accept it as a factor of life and find ways to handle it and to help others handle it. Being a good friend and colleague is listening to others when they are feeling stressed and life is pushing on them.

REFLECTION

Reflection seems to be a lost art today, but I believe it is a very valuable tool. Reflection is the ability to step back and evaluate events, assumptions, core beliefs, and learnings, while drawing connections between different pieces of information. Reflective thinking takes discipline and time, a person needs to make the effort to step back and examine the events of the day, week, or over the course of a project, and use it to learn from those events and the actions taken. Reflection is a way to learn not only from your own experience, but from others. Reading articles on leadership or personal development can help stimulate reflection.

A good way to start is to make a list of questions. The questions can act as the catalyst for your thought process.

- How did I handle the situation?
- Are there other approaches I could take?
- Now that I have wrapped up this project, what are my next steps?
- Where do I want to be in five years?
- What does that article mean to me?
- What impact do I wish to have on my team? Short term? Long term?
- What unique value can I add to my organization?
- Is this important to me? My family? My team?
- What did I learn?

Reflection is a way for you to clear your mind and think clearly about events and the path forward, it is not a quick pace thought process. Reflection is taking the time to think about the day or events and step back so that you can ask the right questions and find calm solutions or answers. Reflection is not meditation, but a deliberate thought process used to help see through the confusion of the day. Self-reflection should become a weekly, if not daily, part of every individual's routine. Besides, daily there are some milestones that should trigger a reflective session including:

- At the end of each major project or milestone
- After giving or receiving a performance review
- After a major presentation
- At the end of a class, go over the major learning events
- After a life changing or major event (death in the family, company merger, breakup).

Reflection is not a time for self-pity, but a time for slow and deliberate thinking and the starting of laying down the strategy and plan for moving forward. Reflection is the time to re-examine where you are and what a set of events or new-found knowledge means to you.

FEEDBACK

We will all get feedback throughout our careers. In many cases, we should seek out feedback from trusted advisers, friends, colleagues, coaches, and mentors more than most of us do. Some of the feedback will be positive, and some of it will hurt! The good thing about feedback is that it is information and potentially insight that we can either decide to use, or after careful consideration, leave on the side of the road of life. Feedback is not a ball and chain or baggage that we have to lug around for the rest of our life.

Receiving feedback that we agree to or puts us in a positive light is easy "I know I'm great." With all feedback, especially with feedback that we don't agree with, the idea is to make sure we examine it thoroughly and not just throw it into the dumpster. We need to think and ask the questions that will lead us to a better understanding of the feedback and the motives of the person who gave us the feedback.

When receiving feedback that we don't agree with, be it a performance review or unsolicited advice from a colleague, don't just think "that is absurd." Take a deep breath and a step back. Don't just think about what's wrong with the statement, but what could be right with the statement and what is the individual really meaning to say. A lot of feedback is given in vague terms "show more initiative," "be more assertive," "show more leadership," "speak up," or "be more creative." You may want to take the feedback and walk away and take some time to think about and reflect on the feedback or ask for a second opinion. When reviewing feedback think of these questions:

- What is really meant here?
- What is the feedback really telling me about me?
- What part of the feedback is correct?
- Do I need clarification?
- Is this feedback clear, concise, logical, and example based?
- Does the person giving the feedback have a motive?

Oftentimes, once we ask ourselves these questions, we find we need to go back and get clarification. What did you mean when you said I needed to "show more leadership"? The person may have just come out of a meeting and knew you had more to contribute in the meeting than you did, and they wanted you to step up and speak up. Taking the time to step back and ask the clarify questions just opened the door for more dialogue and some really good information. The initial feedback was poor, the follow-up feedback is valuable insight that can be worked with and built upon.

At times, we will all get feedback that is negative or just hurts, especially our egos. Negative feedback can at times be great learning and building material, once we get over the initial sting. The key is to listen carefully and not to get defensive or argumentative. There are two key steps to take that will allow you to use negative feedback and to control both the situation when you receive that feedback and how you act upon it.

First, listen carefully. When a person is giving you feedback, let them talk and say what they want to say before you say anything, do not interrupt them. Listen so you understand what is being said, is it fact or opinion? That you left some test results out of the presentation is fact, that the presentation was presented poorly is opinion (at least on the surface).

Second, examine whether the information is accurate. Accuracy and quality of presentation of the feedback are two different things. Most negative feedback seems to be delivered or received in a harsh, insensitive, and callous manner. Few people have developed the skill set to deliver negative feedback in ways that the receiver does not get uptight and defensive. Accuracy pertains to if the feedback is right at least in some portion of it. Remember, even salt is useful in the kitchen.

When examining the information, follow the five actions below to ensure you get the most out of the feedback and use what is beneficial for you.

- *Don't get defensive*: When getting negative feedback, it is human nature to get defensive and argumentative. When the other persons' feedback hurts, seems uninformed, weird, petty, or just off base, it's easy to become defensive or dismissive. When given negative feedback, we tend to look for inaccuracies or holes in the logic, and to then think about a response to refute those inaccuracies. By nature, we want to become defensive and argumentative. We need to get over this urge to be defensive. We need to listen and ask questions, to examine the feedback for the true meaning and message. The key is to listen to the person and let them speak without interrupting, planning a reply, questions, or worse, attacking the individual giving the feedback. Let the individual have their say, hear them out.

- *Ask for clarification, ask non-defensive questions*: After the person is done, take a deep breath, and then ask your questions. If you are not clear about what they said, ask questions to get insight into the feedback. Ask for examples of when you missed the mark, repeat points of what they said, "I want to be sure I understand you," "Do I have it right that you feel that I ..." Asking questions gives you time, can give you clarity of the situation, defuses the situation, and actually puts you in control or at least on an even playing field and promotes cooperation. Questions let the person know that you heard them and are listening.
- *Ask for time*: Asking for time to think about what has been said does several things. First, it gives you breathing room to go off, think about, and digest what has been said in your own time and space. It lets you set a time frame of your choice to get back to the person and discuss the feedback. It defuses the situation by giving you breathing room, and it tells the other person that you valued the feedback enough to want to go off and digest it and get back together at a later time. You can tell the individual something like "thanks for the feedback. I would like to think about this for a few days, and then get back to you." Take the time to think about what was said, the intent, and what it means. Ask for the time, and use it to your advantage, including learning more about yourself.
- *Reflection*: Asking for and taking the time to think about what the feedback is, its accuracy, the source, and how it may be usable is critical. Reflection is a good tool to use in these situations. Take the time to sift through the information and think about it. If necessary, ask others for input and confirmation on your own feelings and thoughts as you go through the reflection process.
- *Action*: Feedback is always something that at the end of the day we can control. We can decide to act on it and in what manner. A lot of the times feedback in the way it is given, be it context or substance, may not be useful, but often times there is a nugget of information or potentially wisdom that we can use and build upon. The choice is, do we want to look for it, and how do we want to act upon it?

Remember feedback helps us grow and improve our skill sets and ultimately understand ourselves better. When given feedback, listen, thank the person who has offered you the feedback, analyze and clarify it, and then decide how to use it. Not acting on feedback is a decision. When getting feedback either when its solicited or not, never shoot the messenger no matter how much you want to. Remember it is up to you how you handle feedback. Often there is something within the feedback that we can work with. Always remember to thank the person who has given you the feedback even if it was not solicited, in this way, they know you have heard them and thought about what they have said. The choice is yours at how you want to utilize the feedback.

Some feedback is petty and of no value other than giving you more insight into the other individual, their insecurities, and motives. Other feedback can have great value and can be used as a tool to grow and understand ourselves, our strengths, and weaknesses better. We can decide how we use feedback and treat it as a gift or something to be left on the side of the road.

DIFFICULT CONVERSATIONS

We will all face difficult conversations at work, those conversations that you know you have to have and spend half the night worrying about the day before. It is human nature to want to avoid these conversations. Since we know we will have these conversations at some point, the key is to learn how to have them in the most productive manner possible. At the end of the day, the end game is to keep the relationship intact and to produce an equitable outcome.

The first step is to change your perspective from "difficult" to a discussion of growth, development, or understanding. When we label something as difficult, we automatically put up walls and shut down collaboration. If we are going into a discussion to give somebody negative feedback on performance, we need to change our mindset to it's a discussion about learning and growth opportunities. The state of mind is half the battle.

Relax and set the stage for the conversation:

> If somebody hits you cold with something that catches you by surprise, and it is not life and death, tell them to meet you in your office or the shop in ten minutes. Giving yourself that time lets you clear your head and think. Take the time so you can take a step back, take a couple of deep breaths, and go for a short walk before they have a chance to unload. By giving yourself some time, you can set the stage for the conversation. Generals are always looking for the most advantages ground to do battle on, this is no different. Taking the time to relax allows you to get you mind in the right frame and to think about the situation. Taking time gives you and the other person time to focus on the issue, they may be coming out of a meeting that got them fired up. Give yourself some time, but not too much, you do not want to seem like your ignoring the individual or the situation.

Plan and use the time to your advantage.

Take the time to think about how you want the conversation to go and the desired outcome. You can, jot down notes or key points you want to cover, but do not script the conversation. Conversations will never follow a script, and you will become frustrated, less effective, and appear insincere. If you try to follow or force a script, you will appear disengaged, which will frustrate you and the other individual(s). Remember, the other person has their own point of view and agenda also. The strategy for these conversations should be very flexible, keep your language simple, clear, and neutral. Keep the conversation based on the facts and do not let it become personnel. You're not there to attack the person or another group of individuals. To help set the stage, ask yourself the following questions:

- What is the desired outcome of this conversation?
- Why are we having this conversation? What is the problem?
- How does the other person view the situation?

If you cannot answer these questions, then recognize that at the start of the conversation, you will have to ask the other individual these questions. Show the person that you want to hear their point of view and understand it. Once you understand their point of view, look for overlap or similarity with yours and use it to build upon. At the end of the day, you want the conversation to be productive for both of you. The conversation may not be pleasant, but use it to help both sides learn and grow. At times, those difficult conversations turn out to be not so bad, when you take the time to think beforehand, and go in with the mind-set that common ground can be found.

Three key concepts of every difficult conversation are acknowledge, give back and post mission reflection.

Acknowledge: Always acknowledge the other persons point of view. A conversation always involves two parties, and they will have their own ideas, pre-conceived notions, and fears. These conversations can often times lead to strained working relationships. It's wise then to come at them from a place of empathy. Consider the other person, their points of view, and feelings. The message may not be pleasant, but you can always deliver that message in a clear, honest, and fair approach. Never use the message to attack the individual. A key thing to remember is to not play the victim, you're giving the news and not the one that will have to live with the potential negative fallout such as a termination of employment or loss of work hours. Always give the other person time to talk. When the other person is talking give them your full attention, don't play with your smart phone or be looking at some report, listen and be part of the conversation. If you listen, you will hear how the person is feeling and help address those feelings and let the conversation get to the point you want it to get to. Listening and empathy may not change the outcome of the conversation, but it will let the other person know you care and that you value the individual as a person.

Give Back, Always Try to Give Back: Often when we have to give somebody bad news, we forget to give something back. When you're thinking about the conversation and getting into the right frame of mind, and the conversation is going to put the other person in a rough spot, think about how you would feel and if there is something you could do to take some of the sting out. Put yourself in their shoes. Are there potential ways you could help, even if the help won't change the final outcome or the conversation? If you are letting somebody go for reasons other than ethical, can you write them a letter of recommendation, provide information on an upcoming job fair, resume advice, or job counseling. If you are telling your boss that you cannot take on a new job assignment, offer alternatives or throw out a colleague's name. Proposing solutions or options shows respect and that your human also.

Post Mission Reflection: After every difficult conversation, take the time to reflect upon the conversation. Did it go as well as you would have liked? How was the news taken? What were your reactions during the conversation? What were the other persons fears and reactions during the conversation? Learn from the conversation and how to do it better. Difficult conversations are never going to be easy, but by learning from each one, we can make the next one better for both parties.

LEARNING

Get into the mind-set that life will always be a learning process, and that you should just plan on a lifetime of learning. Many are tired of studying when they get out of a university and have had it with books. In today's rapidly changing world, lifelong learning is a must. The ability to understand new technologies and learn new skills will be a constant that all engineers and individuals in today's workforce will have to have. Learning is becoming the new norm; many skills and technologies are changing so fast that skills learned today are obsolete within 5 years. Today's workplace requires individuals that are fast and autonomous learners, who spend time and effort staying abreast of industry, technology, and economic shifts.

Learning does not have to be just refreshing a particular skill set, but may include learning new fields, such as engineers learning business skills such as finance or economics. These skills will help the individual make business decisions and help the technical teams be able to close a project's business case or lobby for more resources.

Organizations need individuals that have a desire to stay relevant in today's fast pace and constantly changing business environment, gone are the days when you can slide into a job and do the same thing the same way for the next 30 to 40 years.

There are several ways to stay current in today's business environment. Join a technical society, subscribe to journals, and attend conferences. An individual can go back to school and take classes or get a second degree or certificates. The key is to keep the mind active and learning new things. Research has shown that studying different or divergent fields can help broaden an individual's skill set and increase creativity and innovation. For example, an engineer studying art or business.

Companies also need individuals that are not only consistent learners, but are also willing to share what they learn. The willingness to share ideas, knowledge, and experiences not only builds the knowledge base of the organization, but builds trust and respect among team members. Teaching and sharing of ideas and knowledge also reinforces that knowledge base within the individual.

Organizations can help promote learning by promoting a culture of continuous learning. Individuals should insist that the organizations establish and maintain a culture that supports continuous learning and improvement.

The following are ways that managers and individuals can create an environment of continuous learning and improvement:

- Become a member of a professional society
- Mentor at a local school or university
- Attend conferences or trade shows
- Write papers and give presentations at conferences or trade shows
- Invite vendors in to talk about their products
- Invite distinguished professors or retirees for lunch presentations
- Conduct lunch and learn sessions to discuss best practices
- Invite peers from other organizations to talk about challenges they face and their best practices
- After every business trip, insist on a written report and presentation that covers, purpose of the trip, key points observed, and implications for the organization, business, and industry
- Take classes, these do not have to be in your field. Learning new skills outside our technical field or profession can give us new insight and broaden our knowledge base. Ideas and knowledge from other fields of interest can open up new ways of thinking or new concepts in our primary field of interest
- Do short projects in other areas, engineers should spend time on a marketing campaign, and accountants need to spend time out on the shop floor
- Read, reading stimulates the mind, do it for fun, and to stay current in your own technical field.

If we are not learning new things and new approaches, the pace of technology and the world today will advance past us, we need to stay current not only with technology, but with the world around us.

QUESTIONS

This section could go under learning or with a discussion on conversations, but it deserves its own section. Questions can help you not only understand something, but can help others understand as well, for this reason, questions and the ability to ask questions without fear is a valuable lesson learned. The ability to ask questions is an integral part to the art of having a conversation and to understanding.

Questions are used not only for personal understanding, but can also be used in the case of mentoring and coaching to help others understand and provide options or clarity to problems or a changing landscape. The ability to ask questions is a skill that can be learned. Questions enhance learning and the exchange of ideas, information, and improves innovation and creativity, and can improve organizational, team, and individual performance.

The first step to being able to ask questions is to be a good listener. Nobody likes to be interrogated or asked questions as if they are coming from a robot or off of some checklist. The key is to ask questions that others will like to answer, will entice, or enhance the conversation to keep it moving. When questions are asked, it naturally keeps a conversation moving. Ending a sentence with a period means the end of that sentence and train of thought. Ending a sentence with a question mark naturally moves the conversation on. People naturally like to talk about subjects they know about or are curious about. Asking questions increases learning, dialogue, and personal bonding.

To help keep conversations moving, use follow-up questions. Follow-up questions signal to the other person or group that you are listening, care, want to know more, and show respect for the other person by showing that you are present in the conversation. People interact more with others that ask follow-up questions. Follow-up questions are open-ended questions, they leave room for the person to answer freely unlike closed-ended questions that solicit a specific response such as yes or no or a specific time or date such as Tuesday the 14th.

When in group conversations, understand that group dynamics can drive the conversation. People when in a group tend to follow one another's lead. The presence of others affects the willingness of others to open up, but once the "ice" has been broken conversations tend to flow as long as the collective group does not feel threatened. A good way to handle a group discussion or a brain storming session is to start the conversation out on neutral ground. Ask an open-ended question that either helps set the stage, for example, our customers have given us feedback that the user interface is cumbersome to navigate, the IT team believes that the interface is adequate, where do we think the potential disconnect is? Or ask a question that the whole team can answer and leads into other questions and learning. What was the best vacation you have ever been on? This question can lead into questions about why and how you planned it, that can lead into the interface question. Always break the ice and make the conversation non-threatening.

We learn by asking questions, questions of ourselves, others, and the environment around us. We build bonds, understanding, and creativity by asking questions. Conversations built on questions, thoughtful answers, and respect build rapport both at an individual and team level. Learn the skill of active listening and couple it with the power of questions to be a great communicator. Question everything, questions are the lifeblood of curiosity, innovation, and individual development.

PURPOSE/PASSION

You read and hear about "living with purpose or living with differentiation" a lot today. I will give you my take on the subject based on 30 years of working as an engineer, mentor, coach, and manager. One of the first things I tell those working with me, especially young engineers, is that if you don't like what you're doing then we need to talk, I'm not talking about not liking the day-to-day issues, but truly dreading coming into work then we need to talk. What I want to stress here is find your passion, what truly drives you, and makes you tick. I have known engineers that become engineers because their parents told them to or they thought it was going to be easy for them because they found math easy. But being an engineer did not drive them, and after school and a few years on the job, they hated it. The problem with hating your job is that it will show up in every aspect of your life, the quality of your work, and relationships.

Purpose is not something you find, it is something that you build and develop and at times redefine and rebuild. Purpose at work even in jobs that some may find mundane can have great purpose and personnel satisfaction. Take the taxi driver I meet in New York, "my job is to meet people, get them to where they need to go and help them have a great day," a sales associate at a major retail store, "I help customers find what they need and make selections that they can live with."

Work is work, and some days are just going to be difficult, but if every day is difficult from a mental perspective, then it is time to find another job. Find the place in work that will challenge you and help you grow or meet some other objective. Sometimes that work passion can be what interests us like medicine, space exploration, or designing video games. Most of us need to work to support ourselves and our families. To find that role at work that will help you thrive may require you to do some reflection.

A common misconception that people have is that passion is only one dimension, but like life, it is multidimensional and potentially comes from several roles. Passion for seeing first flights of aircraft may be the work passion. Being a coach at the local school may be a hobby or a passion of passing on the love of a sport. The family may be the center of one's life and a passion onto its own. Our passions are usually like life and made up of multiple facets. We need to acknowledge that purpose, like life, will have several dimensions and depending on the role we are in at the time that purpose will not be the same.

As life changes, purpose and passion will also change. After we master one skill, or we lose a job or face some other challenge in life, we will see that our life purpose and the associated passion for that segment will also change or be redefined. As we move though life, events will shape our feeling of purpose, the feeling of having a shifting purpose is natural and helps us grow as individuals. As one purpose changes, our passions in other areas will shift and help us balance life out until we regain balance again. No single purpose should define us just as no single role can define us.

From a very young age, we have all played different roles, big brother, son, student, and grandson. Each of life's roles had different requirements, roles, and expectations. As we move through life, we will find that requirements, roles, and expectations all change, and as they change so will our passions and purpose.

People forget that in life there are four parts, family, work, community, and self. Each has its own roles and purpose. The key to life is to find the balance between each of these constantly changing segments of life. We as humans are at our best when we have balance in these four areas. Balance does not mean smooth sailing, life will always challenge us in all four areas. Personnel values, beliefs, and friendships can help us find the ability to cope with life and help us find and maintain our purpose.

2 Leading and Working with Your Team

All organizations run on teams. Teams are part of our everyday lives from the school bake sale, church choir, to the manufacturing tiger team at work. A well-run team is always better and delivers more results than individuals could. But a poorly run team wastes time, resources, and in some cases cause individuals to lose faith in other team members and the organization. Teams are a necessity in today's business and technology environments, so learning how to be a good team member and team leader is vital. By understanding how to build a great team, team members can help build the team, learn how to hold team leaders accountable, and ensure the team's success.

In this chapter, we will discuss team dynamics and give insight in how to build and lead a cohesive team. Teams are so commonplace today that we don't give them much thought, especially out on a shop floor. When we think of teams, we think of design groups or groups that are working to develop or bring a new product to market, but the same dynamics are alive on the shop floor. The same practices used to build an effective design team could be used to build a great team on the shop floor or the sales group. Team building skills can be used to build unofficial teams or teams that are not sanctioned by management, but come together by member consent to share information or work on an agreed-to project.

Teams are built around five key principles. The following key principles are not listed in any specific order, but are equally important if the team is to function at peak performance.

TRUST AND MUTUAL ACCOUNTABILITY

A leader must always remember that trust must be earned. Trust, loyalty, and commitment cannot be dictated or coerced, but must be earned and maintained. The members of the team must believe that the team leaders have their best interest at heart and will support them. Team members must also know that each team member will pull their weight. Mutually agreed upon roles, responsibilities, and goals will help team members ensure mutual accountability. Leaders must help team members enforce those roles and responsibilities evenly and thus build team loyalty and trust.

A COMMON SHARED PURPOSE

Teams must be formed to achieve a clear purpose or mission. Many teams are formed or brought together to meet a mandate supplied to them from an outside influence, perhaps the general manager, or to perform a specific task; build wing skins. It is up to the team guided by the team leader to create a team mission statement, one that each team member can believe in and support. The team must know their purpose and take that purpose to heart if the team is to be successful. An effective team purpose or mission statement will have an element of winning, being first, revitalizing, or other rallying point that defines and brings the individuals together as one entity. The mission statement should be simple, easy to understand, and explain and state a goal and reason for being. The cookie sales team will make and sell at least 40 dozen cookies at the school bake sale supporting the senior art day field trip. This statement tells you who the team is, the goal, and the purpose.

People want their work to matter and be important to the organization. Leaders need to recognize that if a "team" is formed, the team must have a purpose, a reason for being. The purpose of the team must require that the team come together and accomplish something that could not be done by the individuals working on their own. For its part, management must hold the team accountable to achieving its purpose and goals. The team leader and management must work together to ensure team accountability to its mission and goals by setting milestone targets and a schedule.

GOALS

The team's goals flow down from the team's mission. While the team's mission usually remains static once agreed upon, team goals can and often are adjusted as needed. A team's goals must be built around the team's mission and purpose statement. It is critical that team goals are understood and accepted by each member of the team if the goals are to be achieved. When talking about goals, two more factors enter into the responsibilities of the team leader. First, make sure that the goals are important to the organization. People work at their best when they know that their work is valued and important. Second, make sure that the goals are reasonable and achievable. Reasonable and achievable does not mean easy, it means that the team has the skills, resources, and support to achieve the goals. The goal may be to bring a product to market in half the time of the last product offering, bring a new assembly line up and into production in nine months, or to reduce line cost by 10% while decreasing defects by 50%. Goals must be kept in sharp focus and become the driving force for the team. Goals need to be simply stated and quantifiable so that team members know when success has been achieved.

SKILLS

When a team is formed, the team members must have the right skills to perform the task at hand or must be given the time and ability to acquire the correct skill sets. When the team is formed or when new members are added they may not have all the necessary skills. For example, a new member may not have the skills to run some of the equipment on the line or the team may find that they need engineering support to modify a piece of equipment. The team members must be allowed to become trained on the equipment or reach out and get the resources they need to be successful. The ability to acquire new skills and training could be a great motivator for team members, while a lack of training opportunities or the inability to become trained or acquire the right skills and resources could be a hindrance to team morale and performance.

TOTAL TEAM

Commitment and understanding as to how to work within the team is accomplished by clear knowledge of individual roles and responsibilities. The team must agree on and know the roles and responsibilities of each team member. Who will do what jobs, the responsibilities of that job, and commit dates are critical information for each team member to know and agree to. Each team member must know who does what and how they are being held accountable. In truly effective teams, all team members contribute to the success of the team and perform real work supporting the team's goals.

In a nutshell, a team is a small group of committed individuals with complementary skills that come together for a common purpose to achieve a set of mutually agreed on objectives. The purpose, goals, roles, and responsibilities of team members are agreed to and supported by each individual team member. Effective teams are based on trust and commitment from each team member. Team members must hold themselves mutually accountable if the team is to succeed.

The five principles of effective teams are based on a set of values that encourage commitment, mutual support, listening, giving others the benefit of the doubt, and recognizing the work and achievements of others. Teamwork is not focused on the individual, but on the collective performance and support of the entire team. Leaders must remember that organizations are made up of individuals and have individual contributors, but teams run on mutual respect, commitment, and trust in others and the holding of others mutually accountable for the team's success or failure.

The five principles help the team form a social contract that will guide the team on its journey and keep it grounded. At the core of the five principles is team commitment and trust. It goes against human nature for an individual to put their fate in the hands of others. By understanding and following the five principle's, commitment, trust, and team unity can be established and maintained.

The cornerstone that all five principles are built upon is good communication. Communication will make or break any team. Many teams with very talented individuals have failed because the team suffered from poor communication skills. The ability of the team members to talk openly and share information and their thoughts is crucial for success.

The most effective teams meet outside of the formal team meeting times. Teams that chat and share information at coffee breaks or lunch build team cohesiveness. Social time builds team unity. Social time gives team members an avenue outside of the formal team meeting to share ideas and information. Social time also gives team members a chance to talk and get to know one another which builds empathy, understanding, and a sense of belonging. Leaders should look for ways to encourage social mingling within the team.

Social time does not mean happy hour, it means providing or encouraging face-to-face exchanges between team members. Leaders need to encourage and even actively set up ways for team members to engage each other. One way to do this is to set up "meetings" outside the standard team meeting time to discuss specific points, for example, if the team is looking at new tooling ideas, go and visit the tooling shop. Another method is to just stop off in the team area and say hello, or reorganize the team at an all-hands meeting, or move the team around so every member is co-located. Creating informal social time will take some thinking, but will add greatly to the team's cohesiveness and success.

Since communication is critical to a team's success, leaders need to pay attention to not only creating the right environment, but also on improving communication skills within the team. Poor communication can destroy a team and become the leading reason why it will fail. Poor communication can lead members to not share information, not voice concerns, and make the team unable to make an informed decision. Poor communication and poor communication skills among members can lead to not only members withholding information, but to attack another's ideas and lead to bad decisions.

Leaders can improve communication skills within the team by following these guidelines:

- *Ensure that everybody has a voice.* If somebody is not talking or giving input in a meeting, call on them for input, ideas, or to voice their concerns. To be committed and to feel part of the team, everybody needs to know that their voice matters and will be listened to. Timid people will talk and share ideas if they feel safe and supported by those around them.
- *Instill personal respect within the team.* If somebody is talking, do not allow interruptions. Ensure that people are given the time to finish what they are saying before they are asked questions, or the next person speaks. Do not allow individuals to bully others or dominate a conversation. Asking a question is not interrupting others as long as the question does not come in the middle of a sentence and is related to the conversation.
- *Focus on behavior, not character or the individual.* Give people a chance to vent and express frustration or anger. But ensure that when people do vent, it does not become a session of character assassination. Focus on the issue and behavior. For instance, if a team member missed a deadline, focus on what the ramifications of them missing the deadline are. Do not say "Because of you being lazy and stupid I had to pull an all-nighter to get the data for the meeting today." Instead "Because I did not have the data yesterday at 4:00 as promised, I had to pull it together myself and create the summary for the leadership review today." By using the "I" word and focusing on the behavior and event, you can open up communication, correct possible bad habits, and enforce commitment. You may find out that the individual had competing assignments, and the discussion will help the team reallocate resources to ensure all bases are covered in the future. If a person feels attacked by other team members, they will end up not being committed to the team.
- *Create rules for team discussions, especially contentious discussions, and enforce them.* Acknowledge to the team that contentious discussions will happen. People will have different viewpoints and ideas and will need to feel

free to express those views. It is not a matter of if, but when contention will happen. By expressing different viewpoints or by questioning another's idea, contention within the team will happen. But the very discussions that can cause contention within the team could lead to the best ideas being formulated or bring about the melding of different solutions into a breakthrough. Contention and tough discussions within the team could, if handled properly, lead to the breakthrough that the team and the organization needs.

- *By setting up rules at the start of the team forming, confrontational situations could be avoided.* When people start to get irritated or withdraw, remind the team of its mission and purpose. Rules such as:
  - "Only one person speaking at a time"
  - "Every idea is a good idea and needs to be heard"
  - "Every idea will get its day in court and be vetted out by the entire team"
  - "Wait for the person to stop talking before jumping in"
  - "Talk in a normal voice, no shouting"
  - "Fully articulate support for or against an idea or view"
  - "All discussion is centered on the idea and issue at hand not the individual"
  - "Contention within the team will happen and can be a good thing if handled properly."

- *The team leader can help use contention to the team's advantage by not only enforcing the team rules, but by also applying the following tools.* Use and encourage active listening. Restate points, ask clarifying questions, refer back to data or previous points made, and ask the speaker to explain reasons behind their stance. All of these actions encourage active listening and encourage team participation.

- *Encourage all views to be heard and vetted.* Use active listening skills to help the team vet all ideas and identify points of contention. A point of contention may be a force or influence outside of the team, such as a policy in one member's home organization. Find the root cause of the contention and look for ways to negotiate a resolution to the issue without insulting or attacking the individual that has brought up the issue. By leveraging active listening skills and ensuring that team rules are followed, a leader (and team members) could often help an individual uncover the reason for the contention and help the team negotiate a resolution. Remember to keep bringing the team back to its mission, goals, and purpose, use these agreed upon foundational elements to unite the team.

- *Team meetings are a major source of open communication and information sharing between team members, hence, it is critical that team meetings are run as effectively as possible.* The following guidelines will help ensure that team meetings are effective:
  - Set an agenda before each meeting. Make sure that the agenda is communicated to each team member before the meeting, especially if they are to present something at the meeting. Allocate time allotments to each subject to ensure that the team stays focused

- Ensure that each member of the team contributes to the discussion or invite them to join in
- Stress that all views need to be heard
- Open each meeting by going over the purpose, objectives, and agenda of the meeting. Stress how the meeting ties into the overall goals and purpose of the team
- Bring closure to each agenda item. Do not let discussions remain open. Open items or viewpoints not mentioned in a meeting will creep into hallway discussions and could lead to tension within the team
- If decisions need to be made in the meeting, drive the discussion so that the team reaches a decision. Use the teams agreed upon method of decision making to make the final discussion (majority rule, consensus, small group, or leader)
- End each meeting with an action plan. Make sure that during the meeting action items are taken. Go over the actions taken in each meeting and make sure that the individual that was given the action agrees to the action item and deadline
- Publish meeting minutes and the action log and date for the next meeting within a timely fashion after the meeting. In the meeting minutes, publish the items discussed and all decisions made. Meeting minutes are a good way to show team progress and to keep as an active record for future meetings and discussions.

Team leaders must:

- Ensure that the team's purpose, goals, and objectives are clearly stated, understood, and supported by all team members
- The team leader must clearly communicate the organizations expectations of the team
- The team leader is the face of the team to management and key external customers. This does not mean they should be the only spokesperson for the team. A good team leader shares the spotlight with others on the team
- The team leader takes responsibility for the team's performance and holds the team members accountable
- The team leader guides the team and ensures that the team is meeting its goals and achieving its purpose
- The team leader shares in the team's burdens and work. Team leaders must share in the real work of the team and not just be a figurehead. Team leaders must pull their own weight and demonstrate that they are vested in the team
- Team leaders must create the social environment for the team. The team's social environment must be built on trust, free and honest communication, collaboration, commitment, and openness. Team leaders must demonstrate each of these values and demand them from each team member
- Effective leaders focus on insuring that each team member understands the team's purpose and goals and builds individual commitment both emotionally and personally to those goals and to the other team members

- Team leaders must remember that people cannot work in isolation. Communication is the key here, and team leaders must set the example. Ensure that all information is received by all members of the team. Ensure that all team members have a voice. Ensure that success and lessons learned are shared
- Recognize individual achievements within the team
- Build team unity by ensuring that all team members have a voice, are included in team discussions, and team events
- Effective team leaders get to know their team members and build bridges for team members to learn about each other. Group lunches, Friday donut coffee breaks, quarterly events, or community service events all help build team unity and commitment. The more you get to know the individual team members and the more empathy leaders show, the better team cohesion will be. Team cohesion will lead to team trust and mutual support.

Leaders need to remember that leading teams is not babysitting. If the team leader is babysitting team members, the team does not have the right members, proper skill set, or team members are not committed to the team's mission, purpose, or understand roles and responsibilities. Leaders need to guide the team to define and develop commitment to the team's purpose and goals, make the work challenging, and keep the team moving forward.

One way to make the work challenging is to find ways to keep individual team members just outside of their comfort zone. Leaders who challenge and help people constantly push the boundaries of their comfort zone can build highly effective teams. But leaders need to remember that pushing one's comfort zone boundaries increases personnel stress, and they must be willing to help that individual with coping with that stress.

Teams are a necessity in todays fast paced work environment. An effective team will always deliver better results than an individual trying to work alone. By following the guidelines presented in this chapter team leaders and members can develop a cohesive team that is focused on results and the mutual support and well being of all team members. Trust, commitment and good communication is the hallmark of all great teams.

3 Virtual Teams, Leading and Working with Virtual Teams

In today's environment with organizations that are multi-national or have operations that span multiple locations and time zones, virtual teams are a necessity. By definition, a virtual team is a group of individuals that work together, but are physically located in dispersed locations from each other. Virtual teams may span different countries, time zones, or even dispersed locations on the same manufacturing site or within the same city.

The effective utilization of technology allows organizations and teams to work from different sites, but still stay connected. Cell phones, laptops, Wi-Fi, text, instant messaging, FaceTime, and shared websites allow individuals to stay connected. But the same technology that allows us to stay connected can make it a challenge to be effective as a team. Individuals could feel isolated or on an island of their own, information sharing could become harder or non-existent, and competing priorities could make virtual teams non-functional. Many companies and individuals find virtual teams are less productive than expected, but are finding them more and more an inescapable means of doing business. This chapter is intended to give some guidelines for creating, managing, and being part of a virtual team.

Being on a virtual team can be taxing. The feeling of isolation, the lack of face-to-face time with colleagues, the lack of direction, and relying on others that you never see for work and information could all lead to frustration and a lack of effectiveness. Being on a virtual team could be a challenge, but leading one could be a bigger challenge. In today's global manufacturing environment, virtual teams are a fact. Leading a virtual team could be done successfully and could lead to greater results than a team that is made up of individuals from the same location.

Virtual teams allow the combining of resources from multiple sites. Design, material testing, tooling development, and manufacturing could go on simultaneously at different locations and around the clock, thus potentially speeding up the cycle time for a project. By reaching out to different sites, different expertise and viewpoints can be leveraged to great advantage.

PEOPLE AND THE VIRTUAL SKILL SET

Teams are successful because they bring the right people together for a purpose. People who work on virtual teams need not only a technical skill set, but a virtual team skill set. The virtual skill set includes good communication skills, the ability to utilize virtual enabling technology, emotional intelligence, the ability to work independently, resilience, the ability to accept when technology and things don't go according to plan, and an appreciation of other cultures and work habits.

Team leaders need to pay attention to how the team is performing and if the team members have not only the technical skills, but the virtual skill set needed to help the team succeed. If a team member has a weakness, be it in handling technology or in one of the other skills, the leader needs to take steps to coach and get the individual the training or help that will improve their skill set, thus ensuring the individuals and the team's success.

In many respects, virtual teams are no different than traditional teams. The same five principles apply to virtual teams that apply to traditional teams:

1. Trust and mutual accountability
2. A common purpose
3. Defined team goals
4. Individual and team technical skills
5. Agreed upon roles and responsibilities.

The principles don't change from a traditional team to a virtual team, but the working environment, the culture, and the way that cultural environment is established and maintained do change. When working in or leading a virtual team, technology can be the team's biggest hindrance or the defining factor in the team's success. The team members' ability to utilize technology will define how well the team performs. In each section below, advice is given on how to use technology to build, define, and run a virtual team effectively.

At the heart of the virtual skill set are communication skills. Some weak communications skills can be overcome in face-to-face interactions. The ability to read body language, mood, and tone all assist in communication, but these factors are missing in most virtual settings. As in traditional teams, good communication is critical, but in virtual teams, it is a defining skill set that all members of the team must excel at.

Communication will be enhanced by the technology that the team decides to use and the rules that they decide to abide by. The tools available to teams today enhance the potential success of virtual teams. Websites, video conferencing, texts, instant messaging, cell phones, FaceTime, and conference calls are some of the tools available today. These tools allow team members to stay connected and able to communicate instantly as problems or questions come up. Task completions and successes can be shared instantly, as well as team or individual recognition.

As the team is formed, the team needs to decide which tools the team will use and how. A resource that almost all virtual teams will need is a good technical resource or IT professional. The IT professional will set up the team's website and help with other technical issues. Even in today's world, many people find technology daunting and difficult to navigate. Once the team is formed, it is wise to have the IT specialist get in touch with each team member and ensure that they can navigate and use the technology toolbox that the team has. It is also wise to have the IT specialist ensure that all team members know that they are there to support them and the team, as needed.

Setting up dry runs to ensure the technology is working at the start of the project is critical for the team's success. If people cannot connect with other team members, they will feel isolated and may even find other work to do. Frustration with technology is a major hindrance and cause for low morale and poor team performance. Technology utilization and ease of use is better today than just a few years ago, but is still recognized as a leading issue and stress point within virtual teams.

Simple text messages and phone calls can help people stay connected and ensure that work is progressing. A team website or "team room" can be a powerful tool. The team room can consist of several walls or folders that contain not only the work of the team, but personnel information such as team pictures and biographies. The team room could also contain a team roster that lists members contact information, location, best communication method, team role and current assignment, and schedule. The team room should also contain the team's purpose statement, goals, and any rules established by the team. The team room could become the coffee house, breakroom, and workroom all rolled into one for a virtual team. Traditional teams have the coffee pot, breakroom, and the ability to see teammates throughout the day, this has advantages for building trust and comradery. Virtual teams need to create that same team comradery and technology provides the team that ability.

Virtual team leaders and members need to use the available technology to stay connected, build trust, empathy, and to ensure that team goals and objectives are achieved. FaceTime and similar technology can tie two or more team mates together for a meeting. The ability to see the other person allows each person to judge body language, build trust, and tie a face to the name. The ability to see the other person makes that person real and not just a name on a roster. If a person is just a name on a roster, it is easy to forget them and to treat them as a non-entity, but once they have a face they become real, along with individual commitments and efforts.

GUIDELINES FOR TECHNOLOGY USE

- *Team Website:* The team's website is the team's own home room. Use the website to keep schedules, meeting minutes, action logs, team contact information, bios, team pictures, team events, and a list of roles and responsibilities and deadlines. The website becomes the team's worksite, breakroom, and conference room all rolled into one. Use the site to bring the team together and connect. Set up the following rules for the team room:
  - Contact information and bios are kept up-to-date
  - Every team member must visit the team's web board and action log every day
  - All data, test results, and progress reports are kept on the site
  - No criticism or offensive material may be posted
  - Set up rules and guidelines for which folders are to be used and content requirements for each
  - Ensure that each team member has a folder on the site to hold their own work

- Ensure that all team members have access
- Have a team page on the site where pictures, vacation information, events, and holidays can all be posted and shared.
- *Team Meetings:* Use video conferencing if possible. If you can see the other person you cannot ignore them, and again, they become real. Some rules to follow when conducting team meetings:
  - Do not allow multi-tasking. Make the rule that team meetings are for team business only
  - Set up meeting times around a common schedule or shift the meeting time around so everybody feels the pain. It is not fair to set a time that is convenient for team members in one location and expect others to call in when it is the middle of the night for them. If necessary, rotate the time or set up a mutually acceptable time
  - Set the rule that everybody must attend the meeting and participate
  - Send out the agenda and meeting information before the meeting and post it to the team's website
  - Send out the meeting minutes and action log after the meeting and post it to the team's website
  - Be specific on the action log. State what the action is and the due date
  - When action item due dates are assigned, keep in mind local holidays and vacations (holidays vary from region to region and country to country)
  - Ensure that everybody has a chance to talk during the meeting. Allow time for any concerns to be brought up and help needed requests
  - To help build trust and for members to get to know each other, start off each meeting with five to ten minutes of personal time. Let members talk about the trip they have coming up, a professional success or achievement, promotion, or their child's school play. While doing this, keep in mind cultural differences, and as the leader you may need to kick things off and set the example. The personal time to some may seem like a waste of valuable time, but it helps team members build relationships and enhance trust. This is a good time to point out upcoming holidays
  - At each meeting ensure that you state the purpose of the meeting and to also recognize individual efforts. Recognizing people can not only help build trust among team members "they are pulling their weight," but also help in turning that individual into a real person
  - Value the team members time, keep meetings on schedule, and if necessary, set up follow on discussion and assign action items as needed. Always strive to keep the meetings to task and on schedule
  - Always listen for frustration, lack of engagement and follow up after the meeting with a call or text message to team members that seem to be not engaged
  - Always encourage open dialogue. Ensure that all voices are heard and issues are vetted out

- Assign a team member the role of "candor." The candor person is to speak up when they think something is being left unsaid or when somebody steps out of line with a comment. The person of candor is to help ensure all issues are brought up and that all criticism is constructive.

PHONE CALLS AND TEXT MESSAGES

The cell phone with the ability to send text messages, FaceTime, and place regular voice calls is a very powerful tool, allowing team members to stay connected. Text messaging has become a favorite method of informal communication. Texting allows for people to send messages to one another around the clock and when the idea or question hits them. Phone calls, especially FaceTime or other visual call methods, are a very powerful tool that adds the visual dimension to the words. FaceTime can be used as a quick review tool, show off a concept or a prototype, and to get instant feedback. But as with any communication method, rules and guidelines must be put in place when the team is formed:

- Respond to all messages within 24 hours, unless on vacation or travel
- Be aware of time zone differences
- If using text, spell the words out. Different abbreviations or XXX may have different meanings in different cultures or to different age groups
- Have the rule that when decisions are made or information is shared that it is posted to the "team room" as soon as possible. If people feel that information is not being shared or decisions are made behind closed doors, team unity and trust can be destroyed.

EMAIL

Email is a good way to communicate to the entire team and to keep a record of what information is being shared. In the age of the laptop, smartphone, and tablets, email is still a powerful tool for keeping teams connected and informed. Keep in mind that email is a type of document that once sent cannot be pulled back. Keep all emails professional and ensure that they go to the right parties. Email is best used to keep the team informed and to share information between team meetings. Email also allows the passing along of documents, photos, and other data quickly. Make the rule that if data is sent via email, that it will also be posted to the "team room."

By posting things to the team room, it ensures that members use the team room, and that all data are available. Emails can inadvertently get deleted, and then individuals have to ask for the data to be sent again. When data are posted in the team room, it is always available. Keep the same basic rules for email that are made for phone calls and text:

- Respond to all messages within 24 hours
- Be aware of time zone differences
- Spell the words out. Different abbreviations or XXX may have different meanings in different cultures or age groups

- Make sure if decisions are made or information is shared, that it is posted to the "team room"
- Advise all team members to reread all emails before sending them
- No non-constructive criticism allowed in emails.

INITIAL TEAM MEETING

As with any team, the initial team meeting is critical. For the virtual team, the initial team meeting can either make or break the team. If possible, the meeting should be face-to-face with all team members and key supporting members in attendance. Supporting members may include key stakeholders, IT specialist, and the team's sponsor or champion. The kick-off meeting will establish the team, and if done properly, set it on a solid foundation.

Supporting team members may not need to be at the entire kick-off meeting, but should be at the initial kick-off session. The team sponsor needs to help the leader establish the need for the team and its charter. The sponsor can erase any doubts about why the team is being formed. Key stakeholders can stress the importance of the team and also layout their expectations. The IT specialist can help the team with any technical issues upfront and help individuals set up their phones or computers with the teams needed software and go over the team's website, access, and structure. Having the key supporting members at the early kick-off session could make the rest of the meeting go smoothly.

The kick-off meeting should build the solid base that will determine the team's success. The kick-off meeting serves two purposes, set up the structure of the team and start building trust and understanding among team members. At the meeting, the team will need to cover the following items. Team introductions, roles and responsibilities, team's purpose, team goals, technology usage and rules, team's website and folder structure, schedule, and work structure. By covering each of these topics and getting the team members agreement to the rules and team operating rhythm upfront, many issues can be avoided later on.

To start the meeting off, the team leader should introduce each team member and give them each a chance to introduce themselves to the rest of the team. When the leader introduces the team member, they should cover why that individual is part of the team and the expertise they bring to the table. By having the leader or the team sponsor do these introductions, it lends credibility to that individual and starts laying the groundwork for trust within the team. After the initial introductions are made, each team member should have a chance to introduce themselves. The personal introductions should include where the person is located, personal notes such as favorite hobbies, family, how long with the company, or favorite vacation spots or trips. By letting the individuals introduce themselves and give personal notes, it helps the team members put a face to a name and turn that name into a real person, again building trust and empathy for the individual.

The initial team meeting will set the stage for how well the team performs. A successful first meeting allows the team to form and bond, which is critical for a team's success. The first meeting also builds trust within the team. Use the first meeting to set the vision and initial goals. Ensure that when the meeting is over, each individual has both a long-term and short-term assignment. The short term builds early success within the team and each individual. Early team and individual successes build trust within the team (that the other team members can deliver) and faith in the team and the teams' mission.

Remember the following when working with or leading a virtual team:

- Early success and trust are a must for any team, especially for virtual teams
- Build trust and follow through on commitments, you don't have the luxury of just walking down the hall and talking to the person
- Keep your communication clear, short, and properly worded
- Use the right communication method for the required message. Use a phone call, text message, or email to cover issues and upcoming personal commitments. Use the team meeting and team room to cover information that the entire team needs
- Always, before you communicate, think what you need to communicate, what results you want, and how the other individual may receive that communication
- As always, but especially in virtual teams, actions speak more than words. If a commitment is made, follow through and hold others accountable
- Virtual teams work and are a necessity in today's global business environment. Virtual teams require work from each team member to maintain cohesion and to accomplish the team's goals.

4 Developing and Delivering Presentations

Presentations are part of every engineer's and manager's life, and giving a great presentation is an art that can be learned. A presentation is given for many reasons: to report data or technical findings, to sell a product or a concept, or for status on a project. There are a few points to remember when preparing and delivering a presentation:

- You can kill an audience with too many PowerPoint slides that just show endless data. Data is great and can be very powerful, but leave the endless data dump for the paper or final report. Hit the high points and the critical message that you want to deliver
- Know your audience, who they are, what they will want to see, and why. Change the presentation to meet the audience needs, even if it is the "same" presentation. A group of engineers and scientists will look at and want different data than a set of business and financial leaders tailor the presentation to the specific audience
- Know why you're giving the presentation: what questions are you trying to answer, and what message do you want to convey?
- People want the presentation to grab them and tell a story, data coming off of slides will put people to sleep
- The audience wants to be persuaded that your solution or the story you're telling is the correct one
- Don't let your nerves kill you or your presentation, stress can be overcome with practice, knowing the data, and believing in the message you're giving.

Let's explore each of these points in more detail.

Engineers love data, and the more data the better, this is true except when giving a presentation. Presentations are best when they are precise, can keep the audience's attention, and gives them a solution or a plan forward. View each presentation as a journey that you're going to walk the audience through. First set the stage for your story. Tell the audience why the presentation is important, why you're giving it, and why they should come along on the journey with you.

People tend to think technical presentations are boring, and usually, they're right because the presenters have not spent the time molding the presentation into a story. The sad part is we, as a community, have come to expect this, and the downside is that we dread endless meetings of presentations that don't grab our attention and, hence, we don't listen to half of it. Managers or leaders often make decisions based on data they only half heard and understood.

Good presenters introduce the topic quickly, tell the audience why they are there, and then present a convincing story, why it's important, and the action required of the audience. Showing an agenda up front does the trick, but it will not grab the audience, the audience wants to know from the very beginning what you're going to tell them and why you're the person to tell the story. People are wired from an early age to listen to stories. An agenda can help set the stage, but it should only be used as a way to slide into the story.

When starting the story, consider the audience and answer three key questions:

- What does the audience already know about the subject?
- What reaction or conclusion do I want the audience to draw or have after the presentation?
- Why am I the right person to give this talk and what have been my (the teams) contributions?

If the audience knows the subject well and you go into too much detail, you will bore them to death. If you go into too much technical jargon, and they have no background, you will lose the audience, and they will start to daydream about what they need to do after you get off the stage. The presenter will have to assume on behalf of the audience on the first question. On the second and third question, if the presenter writes down his answers, it will act as a guide while making the presentation.

When considering the second question, the presenter needs to consider why the audience is there and the information or message he wants to deliver.

If the presentation is of a technical nature, the presenter should make a written document or detailed presentation available before the presentation. This will allow the presenter to focus on the key technical takeaways of the project. If people are interested, they can get into the details by reading the report, which should be made available to the audience if not before the presentation, as soon as possible after it. Engineers love data, feed that need by giving the highlights and the critical data points in the presentation. The key is to tie the data together by using a story. Keep the audience's attention by including the following focus points:

- Why the need for the presentation? Why you're doing the project and why they should care. Draw the audience in by stating that need and the common need to care
- The primary obstacles and setbacks that had to be overcome. Part of the story people love to hear is about overcoming challenges
- The eureka or "AHA" moments. What caused the breakthrough(s)
- Key contributors (no project is done without the help of others), callout key contributors or supporters and give credit where credit is due, share the success story
- Give lessons learned in the detailed paper or the presentation
- Next steps, how the data will be used (cut production cost by 20%) or move on to phase two with the ultimate goal of $5 million in potential sales. Always leave the audience with a clear ending and that the mission was completed or what the next steps will be

- Always leave time, 5 minutes minimum, for questions. Questions will come up during the talk, so leave time to address those questions. When you run through trial runs of the presentation, think of the questions that may come up so you're not caught off guard. If a question comes up that you do not know the answer for, admit you do not know the answer, and promise to get back to the individual. Once promised, carry through on that promise and get back to the individual. If you try to glance over a question or fake an answer the audience will know, and your credibility will suffer
- Keep to allowed time. Monitor your time and stick to it.

If the presenter keeps the presentation focused and to a clear story line, he will maintain the audience's attention and interests, even if the presentation is very technical and fact based.

Never read a presentation word by word off of the slides, the audience can read for themselves or read the report. Instead, use the slides to assist you in making your points and telling the story. Use the slides as talking points and a place to add some graphics or animation that will help you convey the story. Keep the story moving in an orderly fashion using the eight tips above as a guide. Remember to transition from one point to the next and how each point supports the other points.

Keep in mind that audiences do not like presenters that are on an ego trip. Presenters that are full of themselves will cause an audience to shut down and tune out. No project in today's highly technical world is done by a lone individual, give others their due and credit.

Stories are persuasive and move the audience to have a desired reaction. That reaction may be understanding and support for the project or acceptance of the results. Stories motivate people to action. Technical presentations are meant to give the audience insight, knowledge, and a path forward.

A story shows the audience how life changes for an individual or a group of individuals. Stories show struggle and advancing against the odds. All research projects have struggles and challenges. People come together to find a way forward and succeed. Research and presentations are not Hollywood productions nor are they boring subjects that are just facts and data. Plain facts and data presentations will bore people to death. People want to know the objective, the struggle, and the outcome. Even if the presentation is just some status update, the story is that work is progressing, risks have been reduced or understood, and learnings have been made, that it's important work, and you and the team have the desire to move forward.

Caution: Do not put a spin on a presentation. The world is not always rosy. When conducting technical work and research things happen, and experiments do not always turn out as planned. When the audience hears a spin, they will feel like they are being lied to or deceived. Tell the facts and present the data as the story unfolds. Why were you doing the project, what was the expected outcome, what were the lessons learned and present a plan forward. Remember, many novels have more than one book, and the first book always ends with the individual or the team at a crossroads, but they have a plan on how to move forward.

GIVING THE PRESENTATION

Most presenters get butterflies or nervous right before a presentation. I have given many presentations over my career, some have come off well and some not so well, even a veteran can have a bad day. The following are tips to help you give the presentation itself.

Practice the presentation beforehand. You may think you know the material, but getting up in front of people could make most people nervous. Take the time to go over the presentation and your notes. You will feel more confident, and that confidence will show through when you give the presentation and in the question and answer session.

Look for three-to-five faces in the room, especially key stakeholders and supporters. When you select these faces, spread them out throughout the room. As you talk, make a conscious effort to make eye contact with these individuals. Making eye contact draws the audience into the presentation, it also helps reinforce your perceived knowledge of the material. People expect eye contact from the person talking to them and presentations are no exception. If a person does not make eye contact, it gives the perception that they have something to hide. Making eye contact with people in the audience as you scan the audience and talk gives you confidence and adds validity to you and the data.

RELAX

Being told to relax is hard advice to follow, but just clearing your mind before you step on the stage helps. Take a couple of deep breaths before you go on stage. Taking a deep breath does work, it helps slow your brain down and helps you relax. Before the presentation, go to the back of the room or off to the side and tune out for a moment if possible, use this couple of minutes to take those deep breaths.

Nerves are not a weak point, nor are they a disaster. Nerves help us focus and create engagement with the audience. People in the audience know it's hard to get up and give a presentation, and most of them want you to do well.

Do your homework before the presentation. Go over the material and know it, preparation will build your confidence and help you relax. Give the presentation before hand and, if possible, to a colleague and have them ask you questions. Think about any questions you may be asked, especially if things did not go according to plan or you're asking for more time or funds. You cannot control the questions you may be asked, but you can think about potential questions beforehand and come up with answers for them so when asked you can give clear, concise answers. There will always be a surprise question or two, know the material and practice the presentation and answers to potential questions beforehand.

If a question comes up that you do not know an answer to, tell the audience that you don't know the answer, but that you will get it and get back to the person asking the question. Do not bluff your way through, the audience will pick up on the bluff and will then dismiss the entire presentation. If you try to bluff your way through or skip over a question that you do not know an answer for or don't like, your validity and the validity of your presentation will suffer. A question that you do not know an answer to is just an opportunity for you to learn, and then pass those learnings on to others. Set a time frame and get back to the individual asking the question.

MULTIMEDIA

Remember the old saying "a picture is worth a thousand words"? This is true, and a good simulation or short video is worth 10,000 words. If you have pictures, illustrations, or video that can help your presentation come alive use them, work them into the story. Keep the PowerPoint simple and focused on the key takeaways, don't use 20 slides when 10 would do.

Go to the conference room before the presentation, look at how the room is set up, if possible, practice giving the presentation in the room before you go on stage. Make sure you know how to work the projectors and the phone lines, especially if you're responsible for the meeting. Knowing the technical aspects is a critical part of the presentation homework. If you are using links within your presentation, make sure those links work before you go on stage. If a technical issue comes up, it can greatly increase your stress level and the pressure of the presentation.

Remember you are giving the presentation for the audience, you know the material, you know the story. You're there to tell the audience why the story is important, what you (and your team) did, and the path forward. When you take the emphasis off of you, and put the focus on the audience and the need to guide them through the story, the presentation will become more real and will resonate with the audience. Remember, the presentation is not for your benefit, but for the audiences benefit and to help them learn and connect with the project and its importance.

Presentations are a fact of life, and every life and presentation is a story. Look for, develop, and present the story and the presentation will be easier and well received.

5 Understanding Creativity and Innovation

In today's business environment, organizations are under constant pressure to increase shareholder value, while staying competitive in an increasingly global and competitive market. In the drive to increase shareholder value, consideration must be given to cutting operational cost, increasing efficiency, and in developing new products or services. In today's constantly changing and shifting business environment, organizations that are flexible and innovative have the advantage. In this chapter, you will see how culture affects creativity and how leaders and individuals could build an innovative culture and the creative skill sets for themselves and the organization. In this chapter, you will learn what cultural factors affect creativity and simple tools to increase your creative skills. Individuals and organizations that are creative will have an advantage over the competition, be it for the next promotion, being given the next big assignment, or developing the next breakthrough product. In the following text, when I write the word "organization," the word "individual" could be substituted. Individuals build an organization, it is critical that as engineers, scientists, technologists, and technical professionals, we constantly improve our creative skill set.

FACTORS THAT AFFECT CREATIVITY AND INNOVATION WITHIN AN ORGANIZATION

You could find many definitions for the word and concept of innovation; each is correct depending on the perspective of the individual and the specific use of the definition. All of these definitions center around two key elements, taking a new or enabling technology or concept and finding novel approaches for commercializing the idea or technology that customers will find valuable. All great businesses center around the customer and how to give the customer value, and in the process, create capital and growth for the organization. Innovation allows the organization to give the customer new capability, and if implemented properly, improve the bottom line at the same time. One trap that all organizations need to avoid is complacency. Organizations must stay innovative if they are to stay in business in today's rapidly changing and competitive business environment. To be competitive, an organization must constantly improve operations, quality, and introduce new products while staying cost competitive.

Innovation is not necessarily the one-off bright idea that leads to a new product. It may be a novel approach to using current assets or a new marketing approach. Today's great companies have innovation as an element of the company's core DNA. Innovation and the ability to integrate new technology into novel approaches or services must become part of every company's and individual's DNA.

In today's volatile business environment, I would propose that a comprehensive definition of innovation is the application of technology and knowledge in combination with an understanding of the organization and its customers to address customer and organizational needs that bring value to the customer, organization, and stakeholders. In short, innovation is the development and application of knowledge in a manner that will bring value to both the organization and its customers. Innovation that is not of value to the customer or improves the organizations market or operational potential is of little value.

Now that a clear and simple definition of innovation has been established, the next question that must be answered is what factors affect innovation and creativity within the organization and the individuals within that organization. Innovation and a creative mindset must become part of every individual's and organization's DNA, this means innovation is more than a slogan on a banner, it's a passion. Organizations could enable an innovative mindset by:

- Developing an open and trusting environment where ideas could be discussed without fear of labeling or repercussions (label of being a loose cannon or a crack pot)
- Where individuals could test and develop new ideas without fear of failure and negative evaluations based on those failures
- Where lessons learned and best practices are shared openly and actively encouraged
- Senior engineers and scientists are expected to coach, mentor, and share lessons learned
- The seeking of advice and guidance is encouraged, for none of us have all of the answers
- Facilitating the development and career growth of talent
- A learning and inquisitive mindset is encouraged and supported.

People are the heart of any organization and must feel empowered and encouraged to experiment, question, and develop new ideas and concepts. People must feel free to experiment and try out new ideas or concepts, and then share information and results with the understanding that "failure" and the learnings from that failure may move the organization or a project forward. Organizational openness allows the sharing of information and helps build an atmosphere of trust, where departments and individuals share information, fueling the innovation and learning cycle.

Open organizations invite suppliers, partners, and customers into the development cycle, thus adding additional ideas and insight. Customers give insight into what they like and don't like about current products, proposed products, and give insight into the general market. One trap that teams and individuals must not fall into is expecting customers to articulate what the next big breakthrough product or idea is. Most customers will not be able to give specific guidance other than what they like and don't like about a current product. But some customers are knowledgeable enough about a product to be able to point out specific features or options that could improve it. Customer input is always good, for at the end of the day, bringing value to the customers is the name of the game.

Suppliers and partners, on the other hand, could help give more direct input into a design or invention. Suppliers know the current state and could help the team tap into a wider network of potential ideas and solutions. If intellectual property concerns could be addressed and a truly open environment developed, suppliers could bring a great wealth of experience to the table. When combined, the voice of the customer coupled with the insight suppliers and partners provide, a clear future state product statement could be defined.

The days of the stereotyped mad scientist inventor developing the next great gadget in his own workshop independently is over, if it ever did really exist. With today's technology, ideas could flow around the world to dispersed individuals and teams in minutes. Teams could tap into resources from multiple sites and even get customer input from around the world quickly. Teams and individuals at different sites could build and test components or ideas at the same time and share information and results in real time. The world has become a true 24-7 business development machine.

The model below (Figure 5.1) shows the factors that could affect creativity in teams and individuals and how customers and suppliers could be tied into the process.

The factors listed in figure 5.1 are based on research that I have done. Culture is what drives an organization. Culture is what people within the organization do when nobody is watching. Culture is the hidden driving force within all organizations. Culture drives how customers are treated, how partners are engaged, and how the employees feel about the organization, learning, and taking risks.

FIGURE 5.1   Factors that affect individual or team creativity.

A truly innovative organization is an open learning organization. Innovative organizations allow employees to learn, experiment, reflect, and share results, both the good and the bad, on any project or experiment. Employees and their combined talents are the life blood of an innovative organization.

Innovation in today's fast-paced world is a collaborative skill. Customers and partners could bring a different perspective and insight into the marketplace and help identify opportunities and solutions. Collaboration with partners and customers also helps organizations get around the all too common "not invented here" syndrome. In today's business environment, collaboration and multidisciplinary views and skill sets allow organizations to outperform their competition and increase profits and market share.

Collaboration does not mean giving away intellectual property or trade secrets. I have heard several times the excuse "if we share data we give too-much away." A balance could be found. First, scope out areas of common interests. Share data and ideas around the jointly scoped project and jointly own the developed intellectual property and products. Research has shown that less than 50% of the ideas that lead to a new product came from internal teams or research organization. In today's business environment, organizations must find ways to collaborate with partners and customers in order to survive.

In today's business environment, innovation is a must. Innovation is not easy, nor is it impossible. Innovation requires building the right culture and having trained and motivated employees that are willing and allowed to collaborate and test new ideas and approaches. Collaboration is a necessity. Organizations that effectively collaborate will have a competitive advantage. Innovation requires courage, courage to explore, courage to share knowledge, and courage to collaborate. Collaborative efforts must reach across the organization, as well as outside the organization to include partners and customers.

> Organizations that develop the right culture, strategy, and collaboration skills will be able to innovate and will have a competitive edge. Engineers need to always be on the lookout for ways to increase efficiency, reduce cost, and innovate. Innovation is a requirement for engineering staffs in all companies even if it is not spelled out.

The model above has nine factors that affect the creative ability of the individual engineer and hence creativity of the organization. No single factor is more important than the others. Since teams and individuals are dynamic, the importance of the factors will vary in importance. This variation in importance will depend on the individual's needs, skill level, and comfort in their current role and job or program requirements. The importance of the factors will be constantly changing and leaders and organizations as well as the individual engineer will have to recognize and accept that changing landscape.

The following nine elements are found in a healthy innovative culture.

COMMUNICATION

Good communication is critical not only for creativity, but also for personnel growth. Always strive for clear and concise communication. When discussing ideas, giving an update on a project, or just asking questions, always speak and write clearly and use proper grammar. When you do not speak clearly and openly, people may think you're trying to hide something or that you do not know what you're talking about. When asking questions strive to clearly ask the specific question you have and ask only one question at a time. Questions could help open up a discussion and get more ideas out on the table, but only if the questions are clearly asked and relevant to the discussion.

Always keep your communication clear, concise, and genuine. "Please" and "thank you" go a long way even in today's environment, but the overuse of any word or phrase will make the phrase and its meaning useless. After hearing a phrase such as "And I want to thank you for what you do every day to make us great" for the 100th time, it tends to lose the effect if the team does not believe in the person talking. People want to be appreciated not taken for granted or patronized. Remember you could never over communicate, but keep it simple, clear, concise, and authentic.

TRAINING

Training and the ability to learn new skills or expand areas of interest and knowledge are critical for creativity. Creative individuals are constantly asking questions and expanding their knowledge base. Training and the ability to learn helps keep an organization and its people current with the latest trends and technology. Training and experience also help individuals become more innovative by allowing them to draw on a wider knowledge base and by enhancing the individual's skills set. Training could be used both as a way to enhance an individual's skills and as a reward. Innovative people love to learn new things and tend to appreciate new challenges and the chance to learn.

MENTORING

Mentoring is a tool that all true leaders use. Mentoring and coaching is a way to help guide others and help them improve and grow. People learn best by doing and experimenting. A good mentor will help guide an individual by answering and asking questions. Mentors are an advocate for the individual and a resource and sounding board. Mentors will guide and give advice, but will not layout in detail the path forward. A good mentor will help give the individual the tools they need and the latitude to use the tools, and will then let the individual find the right path forward under guidance. Mentoring is not about making or choosing the path for the individual, but about guiding them and then supporting them along the way. Mentors act as a sounding board. Mentors could also help direct the individual to other key resources that can help the individual or team move forward. By asking questions and acting as a resource, mentors could help an individual or team improve their creativity and innovative skill set.

LEADERSHIP

Leadership needs to set the direction by defining the objectives and the mission. Once the mission has been set, leaders need to help clear any roadblocks that may be encountered. In creative organizations, leaders delegate and allow the teams and individuals to determine how to achieve the objects. If you tell somebody what the answer is, you have shut off creativity and any potential innovation, but if you tell them what the objective is and allow them to define the path forward, then you open up creativity and the thought process.

Once leaders define the mission, their role is to help facilitate discussion, provide resources, training, and support to ensure that the team is moving forward. By facilitating discussion and ensuring that the discussion is open, ideas, lessons learned, and potential solutions could flow, leading to creative and optimized solutions. Leadership sets the tone for the organizations and its culture and will help determine by their actions if an organization is creative or not.

EXPERIMENTATION

Innovation requires the ability for experimentation. Experimentation allows for ideas and concepts to be tried out and developed. Experimentation allows for concepts to be developed to see if they will work or should be discarded or modified into a winning product. Many organizations fail at being creative or innovative because they do not allocate individual time, space, or the resources for experimentation. When challenges arise, teams need the time to brainstorm, try out different solutions, and develop the best approach. Teams also need to be able to call in experts or other resources and share ideas and results. A key element of experimentation is the ability, with leadership encouragement, to share data and results in a risk free environment. People must be able to share what went right or wrong with an experiment or a project so that the learnings and best practices could be learned by the entire organization.

The sharing of information could only be done in an organization that truly supports experimentation and has a culture that supports the open/risk free sharing of information and ideas. Team members need to know that they can bring up ideas and share information without fear of ridicule, punishment, or retribution. Experiments that "failed" or did not meet intended goals have led to some of the biggest scientific breakthrough and game changing products because the "failure" and the sharing of that "failure" lead to discussion that paved the way forward. The sharing of ideas and both technical and non-technical data help teams and individuals gain a better understanding of the issues, challenges, and opportunities coupled with the ability to experiment and try new approaches that lead to creative solutions and exciting new products that customers want and need.

STRATEGY

Organizations, teams, and individuals need to develop a strategy. A strategy lays out goals and objectives with an understanding of the competitive landscape and potential risks. Strategies could include both short- and long-term objectives or may only be for the life of a project or for a set length of time. When developing a strategy or setting goals remember the following:

- Make sure that all objectives are clearly stated and easily understood
- Ensure that key stakeholders and team members support and understand the objectives (otherwise it's just words)

- Is there upper leadership or management support (if not, don't count on support and resources)
- Are the strategy objectives achievable (people won't spend their time and energy on something that can't be done)?

A clear strategy and set of goals that people can understand and support will greatly improve the chances of success. A clear defined strategy shows the path forward and sets guidelines. People want their work and efforts to matter. When people understand how their efforts support the larger organization, they will be more engaged and engagement leads to more creative solutions.

CHALLENGING WORK

Studies have shown that challenging work helps stimulate creativity by helping the individual think and learn. If a task is easy or could be done without thinking, a person can become complacent and tune out or become a robot. But if the individual finds the work challenging and engaging or the task requires a new approach, people will tend to accept the challenge and try to develop creative solutions and approaches. People by nature want to find the easiest, simplest way to complete a task. By providing people with challenging and interesting work, an innovative culture can be fostered.

TECHNOLOGY AWARENESS

To be creative, people need the technical skills to match the challenge. In a manufacturing or technical environment, people need to know not only the state of the art, but also what challenges and research are being developed and faced by the industry. Having the ability to go to classes, take training, attend conferences or write journal papers, and belong to professional organizations could greatly enhance an individual's ability to be creative.

RECOGNITION

People need to know that their work and efforts are appreciated. Even in today's environment, "please" and "thank you" when said with meaning goes a long way. No matter the position within an organization, every individual has the ability to recognize teams or individuals that make a difference. Bring in donuts or cookies on Friday or for a review, and say "thanks for the great effort." A T-shirt, mug, or gym bag is another way to say thanks. To be effective, recognition must be timely and sincere. When you are recognizing people or groups, don't forget the support team. The support team could include your IT group or your OA. Many organizations or tiger teams would fail if it was not for the efforts of the stagehands, those people behind the scenes that make sure it all runs well.

The ability to be creative is not just for the few, the skills and attributes to be creative could be developed and enhanced for all team members. Skills such as divergent thinking, problem solving, and critical and flexible thinking could all be learned and developed.

Enabling a Creative Mindset

Creativity skills could be increased and honed with work and practice. The following are approaches that I have found to increase my creativity skills over the years.

Diverse Interests

Increase your areas of interest or sources of information and inspiration. Research has shown that the more activities that an individual has interest in and participates in, the more potential the individual has for creativity. The spark that will trigger creativity may come from a stimulus that is not associated with the current activity. Taking a tour of an art museum or attending a play may trigger the thought that could help solve an engineering problem or supply the inspiration for an architectural design. Research has shown that the ability to solve complex problems increases when you widen the breadth of your knowledge base and sources of stimuli. The engineer that only reads or studies technical manuals and journals will be a good or great engineer when the design needs to be hammered out, but will usually not be the individual to have the creative thought that generates the design or helps the team find a solution by thinking outside of the box. The individuals that have multiple interests and a wide stimuli base that help trigger their interest, curiosity, and form their knowledge base will be able to think outside of the box when required.

Background Noise

When you're stuck and can't get your mind focused, it might be too quiet. Research has shown that a small amount of background noise could increase the creative thought process. When the mind is trying to process information, making the process harder by forcing the brain to work just a little harder to process and work could be a good thing. This does not mean loud, harsh, or multiple distractions that could overload the brain. Put on some music, go work in a common area or a coffee shop. Be careful—too much noise or distractions is not good and will cause you to lose focus.

SEEK OUT ADVICE OR OTHER SOURCES OF INFORMATION

When you come up against a roadblock, seek out others, especially area experts, trusted colleagues, or mentors that will give you honest feedback and ask probing questions. The sharing of information and ideas increases creativity and the number of potential solutions. The process of just having to articulate and state a problem clearly could help you find the hidden key or trigger thoughts and ideas that lead you to a solution or a different approach. At times we get so buried in the work that we cannot see the forest through the trees, and the act of articulating the problem could lead to a path out of the forest.

Make the sharing of information and ideas a two-way street. Openly share ideas and best practices. As you learn and expand your knowledge base, share your new knowledge with others. This sharing of knowledge helps you reinforce and strengthen your knowledge base.

EXERCISE AND HEALTH

When your brain starts to have problems focusing or you feel like you're in a fog, get up and go for a walk or an exercise session. Research has shown that exercise could enhance creativity. The activity does not have to be strenuous, just the act of getting up and pacing or going for a short walk could help get the creative juices moving. As you move, the body is generating energy, and this generation of energy has been shown to help overcome anxiety and fixation thinking, both of which can block original thought and the creative process.

When you're truly being creative, you can accomplish more in a few hours than you could in a few days of mindlessly working or when you're constantly distracted or unable to focus. Your health and the physical area you're working in could have a tremendous impact on your ability to be creative. Keep your workplace tidy and clean. If you're constantly having to shuffle through papers or clutter to find something, that distraction could cause you to lose focus, and that idea you just had may move to the shadows of your mind. Stay healthy and fit. Remember that creativity is quality work and requires a clear, motivated, and clever mind. A simple cold or the flu could cause you to be distracted and unable to think clearly or to concentrate.

BUILD CREATIVE BRIDGES

As you build confidence and expertise in one area, use that confidence and experience to build expertise in another area. The best designers, the most innovative people, have multiple outlets and areas of interest. The ability to leave one area of interest and focus for a short time in another area could help open the brain to new inputs and stimuli which could trigger creative thought. Throughout history, several of the great inventors produced inventions in several different fields. Leonardo da Vinci, Thomas Edison, and the Wright brothers are all examples of inventors that were creative and had interests in several fields. The ability to draw upon experience from several different fields or areas of interest allows an individual to view facts, data, and events from different perspectives and varying experiences.

I have known creative designers that were good musicians, artists, or performed in theater groups. The ability to mix experiences or to have outlets in other fields, areas of interest, and inspiration allows you to connect disjointed ideas and concepts and think outside of the box. When faced with a design or mental roadblock, the ability to go pick up an instrument or cook a gourmet dinner causes the brain to shift gears and views events and facts from a different perspective.

The ideas and inspiration that spark creativity could arise from anywhere and oftentimes comes from areas not directly related to the field that the creative thought is applied too. When your mind becomes blocked, get up and go do a non-related activity to give your mind time to refresh and focus on something else. The distraction of working in another area of interest helps the brain relax. Creative thoughts usually are not the bolt of lightning from the sky, but instead start out as a random thought or idea that is then built upon.

BUILD EXPERTISE

The adaptation of creativity and the ability to see a need is based on experience and expertise. Knowledge does not hamper creativity. When coupled with curiosity and a willingness to explore and question, individuals with deep routed knowledge in a field will see connections that others will miss. The ability to rearrange, order, and develop something that is original requires the expertise in the field to see what is current and understand what is missing and how to fill that void. If you know the rules and principles, you will understand how the rules can be bent and modified and, in some cases, broken and replaced with something new and novel.

Many believe that a deep understanding of a subject is not conducive to creativity or innovation. This can be true if that deep understanding is not tempered with curiosity and the willingness to ask questions. A deep knowledge and understanding of a subject can blind the individual to the element of what is possible.

KEEP A NOTEPAD

Creative ideas come at any time and often start as a random thought. Keep a notepad on your desk at home and the office where you can jot down ideas or sketch something out when it pops into your head. The best ideas and thoughts are totally unexpected and can come at any time, so it is critical to keep notepads around to just catch the essence of an idea before it disappears like a puff of smoke.

Research has shown that the brain at times will play back the events of the day, especially trying events. The brain can and does work through problems before and during sleep. It is critical to try to capture these events. Have you ever woken with an idea and think you will remember it later just to have it disappear into the morning fog? Ideas are fleeting and must be captured, analyzed, and usually combined with other ideas to develop an integrated creative concept.

CHARACTERISTICS OF CREATIVE INDIVIDUALS

- Flexible: Creative individuals are flexible and will view challenges and problems from multiple angles and will apply different approaches to find solutions
- Not tied to convention: Creative individuals can think outside the box and will look past the normal trends and ways of thinking and explore new approaches
- Curiosity: Curiosity is a key element of all creative individuals. People who are curious will ask questions and explore. The need to learn and ask why leads to not only a deeper understanding of a subject, but also the willingness to explore, ask why not, and look for a better way.

Creativity is expected of today's technical professionals. The ability to combine data and concepts from different sources and areas will help you find that creative solution. The key to being creative is to keep asking questions and keep learning, especially in multiple disciplines or areas of interest.

6 Critical Thinking Skills

For years, we have heard about the need for developing critical thinking skills, but few of us have had any training or even know a good solid definition for what critical thinking is when asked. Critical thinking is the ability or skill to analyze a situation, facts, data, and the way we are thinking, analyzing, and processing data to enable the recognition and filtering of biases, personnel feelings, and opinions so that only the facts are processed and utilized to make decisions, while maintaining an objective position or view on the subject.

Critical thinking is a skill set that must be developed, especially in today's instant information, electronically connected society. Critical thinking enables the individual to analyze facts, evidence, and arguments for validity, truth, and incorrect or illogical elements. Critical thinking also enables the individual to then use the data and evidence to present strong, sound bases for their own thoughts and ideas. Critical thinkers are able to evaluate and synthesize information from multiple sources independently to develop their own logical data driven solutions and answers.

CRITICAL THINKERS

- Actively look at all sides of the argument objectively
- Look for and are open to new views and potentially views that do not agree with your current position
- Review facts and evidence impartially and cross reference the data
- Review all assumptions and evidence for validity
- Research and study independently
- Consult others for clarity, data, references, and opinions even if those views will not support the status quo
- Actively look for biases in the arguments and filter biases out
- Understand the danger of and looks for rationalization disguised as reasoning
- Test the soundness of all claims and presented facts
- Formulates solutions only based on the facts that could be supported after analysis
- Can organize their thoughts and present solutions concisely and coherently
- Attempt to understand and anticipate potential consequences of all solutions
- Is aware and sensitive to the human quality that intensity of a belief adds validity to that belief (if I believe enough then it's true)
- Constantly question one's own views and beliefs and try to understand the assumptions and the implications of them
- Is aware that one's own knowledge and understanding is limited and can be improved if an open and inquisitive mind is present.

Currently, in most coursework, students are taught to regurgitate facts, figures, and formulas. In the real world, we need to look at data and perceptions critically to form answers to very complex issues. Critical thinking is an applied skill that needs to become a habit. Critical thinking is a process where the individual analyzes the argument, data, biases, and their own thought process to develop and evaluate potential solutions. The learning of facts and figures is needed, but the ability to look past the regurgitated data and see what data support the question and why, while considering biases and all qualifications and limitations of the data is a skill that is invaluable.

Critical thinking is a method to look at situations and data, to analyze them to develop unique solutions that are free of biases and could be logically supported and defended by the facts and data available at that particular point in time. Critical thinkers will think for themselves by asking questions, analyzing the data, questioning all biases, status quo, standard beliefs, and then propose new solutions, test those solutions, modifying the solutions as needed, and present a unique data driven solution. Critical thinking is a four step continuous process as shown in Figure 6.1 until a logical data driven solution is found.

In the end, critical thinking is a self-directed disciplined process for thinking and acting. Critical thinking is also a self-correcting process, where all facts, data, assumptions, biases, and first thoughts are questioned until a solid solution can be developed, and then logically defended and articulated. A critical thinker does not take anything at face value, a critical thinker spends time asking questions and thinking about the answers that are received and conducting the follow-up investigation until a solid set of answers are developed and can be turned into a solution and an action plan.

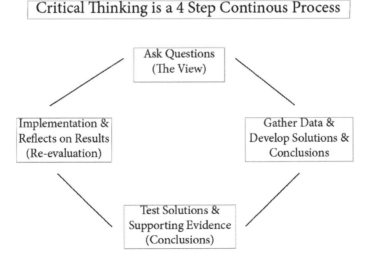

FIGURE 6.1   Critical thinking process.

Critical thinking can be broken down into four steps. The question, gathering data and developing solutions, testing solutions and supporting evidence, and step four, implementation and process review. Specific actions are taken in each of the four steps.

Step 1—The Question(s): What are you looking for and why? What are some sources of information and how reliable are those sources of information? A critical thinker never shies away from questions. Questions can make us feel stupid or foolish, but questions help us understand a subject and fill holes within our knowledge base. The asking of questions is how we learn and grow. The ability to ask questions and receive feedback protects us from group think, biases, uninformed assumptions, and incorrect logic. Develop the question carefully, so that it is specific and clearly stated.

Step 2—Gather Data and Develop Solutions: Ask the experts in the field and do your own research. Cross reference all data to ensure that you have multiple perspectives and sources. Do not rely on a single source even if that source is the leading expert in the field. A critical thinker never uses only a single source of information to base a solution upon.

A critical thinker does their own research. With today's connected society, information is readily available, but this means that false information is out there and must be avoided. A critical thinker questions the motive of all sources and cross references all data and claims to ensure validity.

A critics thinker will never assume that they are right or have all the answers. A critical thinker will always take in other perspectives and points of view, even if those views are contrary to traditional or standard opinions or thoughts. A critical thinker views a subject from multiple angles and viewpoints, accepting that they do not have all the answers, and that their original opinions may not be correct once more facts and data are gathered.

Once data and "facts" are being applied to the original question, it is very common for more questions to be generated and more research required. Once a solid set of information has been gathered, then it is time to formulate a solution or an answer to the question. Use the data, facts, and information to formulate a clear solution that answers the question.

Be wary of biases and data that do not seem to fit. We all have biases, and these biases can cloud our judgment. Data that we bias or treat specially can lead us astray and cause us to misinterpret other data correctly and objectively. Always question the data, both their source and validity. Critical thinkers seek out all sides of the argument and will test the soundness of all claims and data before using that data.

Step 3—Test Solutions and Implement Final Solution: Once a solution has been developed, it must be tested before its implemented. First, see if the solution answers the original question. Second, check and see if the supporting data validate the solution. If the data does validate the solution, then move on to the third review point, if not, look for more data or develop a different solution based upon valid data. Third, consider the implications of that solution. What are the ramifications if I go through with this particular solution? Are there potentially other solutions that I have not considered? Are there long-term effects of this solution, does it impact others, and how? We need to remember Newton's third law, "for every action there is an equal and opposite reaction."

Step 4—Implications and Reflection: Once a solution that is supported and validated by the data has been developed, it is time for implementation. Ensure that you have support from all affected parties, and that they understand the process you went through and the facts and data that support the decision. As part of the process, a critical thinker will consider how the plan is to be implemented effectively.

Don't be afraid to change course if things change and you need to revise your strategy. As more information becomes available, you may find that your question, your analysis and data gathering, or execution need to be refined or totally changed. Life is a learning event, and critical thinkers will be able to shift as more data become available.

After a major decision has been made and implemented, a critical thinker will take the time to examine if the solution was the correct one and if all expectations were met. A critical thinker will also examine if the solution has ramifications or consequences that had not been thought of at the time of implementing the solution. A critical thinker will also take the time to review the entire process to see how their thinking and analyzation process could be improved. A critical thinker always looks for ways to improve themselves and their abilities.

Table 6.1 goes into more detail about each step and different elements in each of the four steps.

TABLE 6.1
Critical Thinking Process in Review

4 Steps and 12 Elements

Steps	Elements of the Thought Process	Goals	Questions?
1	The Question?	Define the goal as a question or a problem statement	• What is the single question • Why the question • What drives the question
1	The Point of View	Define the initial point of view	• What is happening • What are the multiple points of view • Are all views legal and ethical • Who are the players

(Continued)

TABLE 6.1 (*Continued*)
Critical Thinking Process in Review

4 Steps and 12 Elements

Steps	Elements of the Thought Process	Goals	Questions?
2	Assumptions	Gather data for multiple sources	• Look at both sides • Are they reliable • Can the sources be validated
2	Information	Define the initial assumptions	• What are my assumptions • What are my biases • What are the environmental factors
2	Examine the Data	Sort and rank data looking for overlap and holes	• How does the data relate to the problem • How does the data influence or relate to the questions • How does this help us in the data replacement
2	Concepts	Turn the data into viable concepts	• Do the data points fit into the puzzle • What answer does the data point to • Are the concepts feasible
3	Review Supporting Evidence	Review all evidence and data to flush out and narrow down concepts and solutions	• Are we missing needed data • Do we make the facts and data right • Can we explain our biases and assumptions • Can we review the evidence from multiple views
3	Test Solutions & Conclusions	Test viable solutions	• Can we trust the claims • How confident are we • What are the specific findings • Can we validate all claims and tests
3	Re-examine Data	Review test results	• Does any of the data contradict each other • Are there holes in the conclusions • How strong is the evidence and data
3	Finalize Solution	Develop final solution based on available data	• Can the solution be clearly stated • Can the solution be validated • Does the data point to one clear solution • How flexible is our solution
4	Implementation	Implement final solution	• Are we missing something • Are we truly answering the questions and problem statement • Is the solution robust
4	Reflect on Results	Review results and reflect on the process and results	• Did we miss something • How complete was the process • Did our biases or assumptions sway the solution • Did we answer the original question • Where do we go from here

WAYS TO IMPROVE CRITICAL THINKING SKILLS

Critical thinking is not easy, it requires work and a concise effort. The good news is that we can improve our critical thinking skills. The following is a list of activities that can improve our critical thinking skills.

LIFELONG LEARNING

Knowledge is power, and we should strive to learn something new every day. When we come across something new and interesting, we should take a few moments to learn something about it. Lifelong learning is a hallmark of critical thinkers. This does not mean you have to learn quantum mechanics in a day, it means that you should learn something that you did not know before, a fact, an event in history, or something directly related to your field of study. Striving to learn something new every day helps us stay curious and also broadens our knowledge base. Research has shown that learning new information or skills helps form new neural networks which promote brain health.

Learn new hobbies or take a class. Here again, the class does not have to be science related, it can be an art class or an exercise class. The motion of engaging the brain in a new activity broadens our knowledge base and our ability to connect abstract data together into coherent thoughts. Hobbies and the arts help us relax and clear the mind so when we need to focus we can. The act of learning new things helps refresh our minds and helps us view the world through different lenses and from different perspectives.

DEVELOP A QUESTIONING MIND

A two-year-old learns by always asking questions. In lean training, we learned that by asking why five times helps us find the hidden answer or the root cause of an event or action. A two-year-old asks why five times so they can truly understand something and fit it into their knowledge base effectively. Questions drives learning, and learning/knowledge drives informed decisions and actions.

Questions can help us clarify our thoughts, avoid misunderstandings, and remove the fog of uncertainty. Questions can also help us understand and control our own biases and understand those of others. A lot of fear is born out of ignorance, which can be overcome by asking questions. Research and exploration are just activities that are based on questions. The questioning mind is always curious and amazed by the world and people around us.

DEVELOP ACTIVE LISTENING SKILLS

Active listening is a necessary skill for all critical thinkers. People tend to be poor listeners. People have the tendency when "listening" to be formulating responses to what is being said or thinking about another subject entirely. Active listening is very hard work. Actively listening is the act of not only hearing the words being said, but trying to understand the intention, feeling, and message of the words from the perspective of the speaker. An active listener has empathy for the speaker. This does not mean agreement or sympathy, just the courtesy of hearing the speaker from their point of view without initial judgment.

Like any other skill, active listening skills can be improved with practice. The following activities can help improve active listening skills:

- *Talk Less*: Active listeners want to hear what the other individual has to say. You learn more by hearing others talk than you do hearing your own voice. It is impossible to hear and talk at the same time, so active listeners stop talking and let the other person talk

- *Focus on the Speaker*: Active listening is not about you, it's about the speaker and what they are trying to say. Use eye contact to tell the speaker you're engaged. Avoid distractions by silencing the phone and try to block out background noise or interruptions. Make the speaker comfortable and feel like you're engaged by nodding and making gestures that reassure the individual that you are hearing them. Try to make the speaker comfortable in the situation, sit down with them, get at eye level with them, or offer them a drink and allow them time to collect their thoughts

- *Empathize*: View the speaker as an individual that has a story to tell or a point to make. Listen without prejudice or letting biases enter into your thinking. The ability to set yourself aside and listen to another tell their story without injecting yourself into the story is a true skill. When we listen with empathy, we are trying to hear the meaning and understand the feeling of what the other individual is saying

- *Silence*: Silence is not a bad thing. Silence gives people time to think, compose themselves. Silence gives a speaker time to compose themselves and their thoughts. Silence can also give a speaker time to take a drink or a deep breath and continue their story. Silence can give the listener time to reflect on what has been said and the opportunity, if appropriate, to ask a question for clarification or to help the speaker progress through the story. Good communication is the exchanging of information, silence can help that process by being the bridge between segments, silence is the breath at the end of a paragraph

- *Body Language*: People communicate as much or more through body language as they do through words. Body language can give you an indication of the individuals mood and the focus of the discussion or the story. Body language can help you understand the tone of the words and hidden meanings of the story, and if you are getting the full story or just a snippet of a larger story. The eyes and facial expression are a major indicator of body language. Eye contact is a major player in being able to judge the other person. Remember the saying "the eyes are the gateway to the soul," a person's eyes coupled with facial expression will usually tell you the mood that the person is in.

Remember an active listener is engaged in listening to the individual tell their story. While the other person is telling you their story, they are the center of the conversation. Try to stay focused on listening and not let the human nature tendency of formulating a response creep in. Often times when others are talking, we are formulating our response, this is not active listening. Studies have shown that most people can at most only remember 50% of a conversation, and that after 24 hours this drops below 25% at best. To be a critical thinker you must be able to listen to other ideas, plans, dreams, and arguments without injecting your own thoughts until you hear their story.

EVERYDAY SKILLS

Critical thinking can be used in everyday activities such as reading, taking notes, in developing a position or an idea, and in arguing a case or a position. Critical thinking skills are transferable to all of these activities. Well-developed critical thinking skills are useful anytime information must be analyzed and ideas or concepts must be developed and implemented.

Critical thinking allows an individual to read with a critical eye. As a critical thinker, you will ask questions as you read material:

- What is the author's intent?
- What claims or argument is the author making?
- What are the sources of the author's information?
- Are the sources valid and reliable?
- What are the biases of the author and their sources?
- What are the implications of the material?
- How does the material affect me or the world around me?
- What are the counterarguments to the author's position?
- Are the conclusions the author draws sound and valid?
- How can I learn more on the subject?

Reading critically means that you are taking an active, reflective, and enquiry-based approach to reading and understanding the material. This does not mean that everything you read requires your critical thinking skills, some reading is meant for fun and should stay that way. Reading with a critical eye requires work. Critical reading skills allow an individual to see the biases of the argument and data and allow the reader to step through and break down an argument into its smallest elements, and then reassemble the argument and test it for validity. In today's information-heavy environment, nothing should be taken at face value, especially if it is being used to make critical decisions.

Writing is another area where a critical thinking skill set can be very beneficial. Critical thinking skills enable an individual to seek out all sides of an argument and test the soundness of the independent positions. Critical thinking skills also require that all hypotheses need to be tested and validated, thus growing the knowledge base.

Critical thinking provides an individual with a logical approach to not only interpret data, but to present them as well. Writing with a critical voice will help ensure that you answer the question that is being asked and that your facts are accurate and that your argument is sound. When writing critically, use the following rules:

- Read the question several times and break it down into bite-size pieces
- Plan out your response, especially when conducting your researching. Think before you write. Plan out your response and why that response is important to you
- When doing your research, keep notes and also include the sources of the data you plan to use to support your argument or position
- If you do not understand the meaning of a word, look it up, do not assume that the person used it correctly
- Do not use slang or abbreviations in your writing
- Develop a story or a theme and walk the reader through the story in a logical, data-supported manner to a conclusion that the data and research support. Ensure you tell the reader what you want them to know or do and why
- Link thoughts together by using words like "equally," "important," "for this reason," "however," and "beware." This approach helps the reader develop the same conclusion you did and take the action that you as the writer want taken
- Write using proper grammar and sentence structure
- Give yourself time to review your work after you have written it
- Refer back to the question regularly to ensure you're staying on track
- Write clearly and keep it as brief as possible. People will lose interest in a rambling paper.

Writing is a very powerful way to convey a message. Not only is writing a powerful tool, a written document is a permanent record. Remember words in the written form last forever, so make your work stand out and be accurate. When you write, you have the ability to set the stage for the reader, and then walk them through your argument.

In essence, critical thinking skills are transferable to many aspects of life. Critical thinkers strive to be clear in defining the question that needs to be asked, gather data and insight from multiple sources, and then sift through the data with a critical eye. Critical thinkers interrogate all data and sources for correctness and validity. Critical thinkers strive to be concise, logical, unbiased, and fair.

Critical thinking is important because poor thought processes lead to poor and costly decisions, both from a financial and quality of life standpoint. Critical thinking is a process that allows for the development of a question and a solution that is bias free and supported by a logical argument and data.

The critical thinking process requires the individual to stop repeatedly and question their own biases through a process. The process is a self-directed, self-disciplined process that puts the burden on the individual to stop and self-monitor themselves and admit when their logic is wrong, and bias is entering into the process, be it the development of the question, research, validating or cross referencing data and sources, or developing the final argument and solution. Critical thinking requires reflection and open honesty on the part of the individual. This open honesty with oneself may be why so few practice critical thinking regularly. The reflection process does not need to be hard, it can be as simple as stepping back and asking,

- How am I doing?
- Am I being fair and unbiased?
- Have I missed anything, and if so, what?
- Am I being unemotional and logical?
- Have I listened to both sides of the argument?
- Have I defined a clear path forward that answers the original question?
- How can I improve my own thought process?

Critical thinking is the best way to reach a logical answer that is based on truth and supported by logical reasoning. Critical thinking tries to see through the fog of bias thought and develop logical solutions, this is not easy for humans, and even with critical thinking, we may fail, but without it, we are doomed to have biased-based costly and non-optimized solutions.

Please take the time and effort to become critical thinkers in your personal and professional lives. Become inquisitive, judicious, open-minded, logical, and analytical. Don't let others think for you, but question them and their own motives and develop your own solutions. Learn to learn and think for yourself.

7 Developing Requirements, Goals, and Objectives

As engineers, we live by requirements, goals, and objectives, but these words are often misunderstood and misused. Requirements are necessary to define the project's scope and provide specific success criteria. Good requirements help the engineering team and management avoid scope creep and define the project, project success, and develop the progression plan. A well written requirement list is the success criteria for a project.

In most college classes, the subject of requirements and requirements writing is often times not covered or just glossed over. This lack of proper education in the development, writing, and analyzing of requirements could lead to sub-par designs, program over runs in both cost and schedule, and frustration with both design teams and customers. Most engineers and program managers find the writing of requirements challenging.

Objectives, goals, and requirements are all separate elements. Goals are high level and capture the intent of the project, for example, capture 40% of the market share, develop a web-based system that will track investment portfolios under $100K. Objectives are the items that the project will directly develop or accomplish and support the high-level program goals. For instance, the aircraft will fly 7000 nautical miles and hold 240 passengers in a single configuration. Requirements are the specific individual items that must be met to fulfill all of the programs goals and objectives. Each level of the objectives, goals, and requirements list gets more and more specific. Requirements are very specific and single focused. Requirements are the items that must be met, whereas objects and goals are the items that the team should try to meet and are used to give the project scope and overall direction. At no stage should the developer of the objectives, goals, or requirements dictate a design solution.

At the start of all programs, it is critical that the team sit down and develop a requirements document. In some companies, the document is called the technology needs, requirements, and objectives document, in others, it is simply called the objects and requirements document. In all cases, the requirements document lays out why the project is needed, key attributes of the product, and puts the project into context when discussing schedule and budget.

At the start of developing the requirements document, the team should make sure they understand the following and incorporate them into the document to ensure that as requirements are being developed, the team understands the scope of the project fully:

- Needs
- Goals

- Objectives
- Program constraints
- Project purpose/vision
- Budget
- Schedule
- Management/organization structure
- Key customers and stakeholders
- Assumptions (including competition)
- The base line or starting point (if a derivative or a direct competitor to an existing product).

Requirements are developed using several different sources. Teams should get input from internal and external customers and end users of the product. Regulations and business practices should be assessed for requirements application. Market and business assessments and surveys could also be used to gain insight and potential requirements. The competition, technology trends, developmental research, and internal business and research could lead to specific requirements. Good requirements and the research that goes into developing those requirements will lead to better products that customers are willing to pay for.

Research has shown that less than 20% of all technical projects are considered successful, over 30% of all projects are considered failures, and around 50% of all projects are considered partially successful. The lack of complete, clear, and robust requirements up front has been identified as a leading issue with a technical program's success or failure. Missing requirements, vague requirements, scope creep, conflicting requirements, and poorly written requirements documents have been identified as leading causes for program failures and customer dissatisfaction.

Successful programs have been defined as programs that finished on time, on budget, and met or exceeded all of the customer requirements. Failed programs meet none of the three criteria, and partially successful programs either did not meet the budget or schedule requirements or failed to deliver to the customers what was required by the customer at the product introduction. A well written and complete requirements document sets a framework and establishes the success criteria and the basic statement of work for a project up front.

Once a requirements document has been written or even outlined, the document should be reviewed both formally and informally with subject matter experts, customers, end users, and program representatives to ensure that all of the needed technical requirements, timeline, budget, objectives, and the proper scope of the project have all been accounted for. The work done up front in establishing a good requirements document and ensuring it contains the proper requirements, and that those requirements are clear, concise, and could be validated will pay dividends throughout the program.

Writing good requirements is more of an art form than a science. There is no Word program or application for writing requirements. Good requirements writing and evaluation come from experience, practice, and following established rules and guidelines. The following guidelines have been established by experience and are a good starting point.

A well written requirements document can help the team avoid the following pitfalls:

- Writing requirements that will not meet customer needs, are vague, or conflicting
- Developing a program/project that is lacking the needed requirements
- Scope creep
- Developing a program/project that does not have a set budget or schedule in place
- Working to constantly changing or adjusting requirements
- Misunderstandings between leadership, customers, support organizations, and the development team as to what the goals, needs, and the "real-requirements" of the project are.

A well written, solid requirements document serves as the team's guide during the development, testing, marketing, and deployment of any new product. It is easier, simpler, and less stressful to develop a product when you know the requirements and success criteria up front. Also, customers and management will have more confidence in the team, when the team can articulate clearly what the product will do and what they can expect from the new product. A good requirements document provides the framework and overview for the project.

Key words to keep in mind when writing requirements:

- Shall = requirement
- Will = purpose statement
- Should = goal

KEY POINTS

A requirement states what must be done and never states how the task is to be accomplished or a design, manufacturing solution.

Well written requirements do not contain words that are ambiguous or have double meanings. Requirements cannot contain words or statements that cannot be verified or quantitatively defined. The following list of words should not be used when writing requirements:

- Minimize
- Maximize
- Easy
- Rapid
- User-Friendly
- Sufficient
- Quick
- Small
- Large
- Flexible
- Accommodate

- Robust
- Field able
- Extreme
- Variable
- Etc.

Good requirements are written so that the reader knows what must be done and who must do the work. For example:

- The system shall provide XXXX of electrical power
- The complete system shall weigh less than 6.5 pounds
- The system shall be capable of 9 hours of continuous operation at 165F without failure
- The system shall withstand 9G force without failure
- The fuel tanks shall provide enough fuel storage for 5 hours at max engine thrust with 45 minutes of reserve fuel after landing
- The structure shall be capable of withstanding +9G, –2G forces without damage.

VALIDATION

Validation is a key element of all requirements. All requirements must be verifiable by test, analysis, examination, inspection, or demonstration. If a requirement cannot be verified by some quantifiable means, it is not a true requirement. Requirements must be achievable, if a requirement is not obtainable, it is not worth the effort of writing. Achievable does not mean easy, achievable means that there is a potential path forward that the team can explore. Reach targets/requirements are acceptable within requirements documents, but must follow the same guidelines as all other requirements. All requirements should go through a validation check after being written.

The following questions can act as the starting point for a validation check:

- Is the requirement necessary?
- Is the requirement properly written? Is the requirement in proper English, spelling, punctuation, and is it clearly written and understandable?
- Does the requirement have a shall in it?
- Requirements are singular statements. A requirement can have only one who and what
- Requirements cannot have "and or" statements, requirements are specific and state a singular who and the what very specifically
- Does the requirement contradict another requirement? If requirements contradict themselves, you must rewrite the requirements, delete one or the other, or possibly both, and make a single very specific requirement. A requirement answers a single shall statement
- Does the requirement contain specific success criteria, weight, power, and operating temperature? Does the shall statement have stated success criteria that can be validated?

- Is the requirement technically feasible or possible? If the requirement is not technically feasible, it is a research project or a dream and not a program requirement
- Is the requirement written in a positive tone? Requirements should not contain a "shall not." A well-written requirements document will contain the requirements needed to meet all safety, mission profile, system, and structural needs. A shall not, by design, limits the engineers design freedom
- Is the requirement left open? No requirement can have a "to be determined" or "to be resolved" in the requirement statement. You can either achieve success and know what needs to be done or you do not have a requirement. Open issues must be answered before requirements released.

TIPS FOR MANAGING A REQUIREMENTS DOCUMENT

Once the initial requirements document has been written and signed off on, the team must be able to modify it. Change is inevitable during the life of a program. It is critical that the team hold document reviews to ensure that the current list of requirements is still valid. A historical review will turn up many products that were brought to market just to find that the market had changed and the product was no longer what the customer or the market needed or wanted:

- The team must maintain a change record for the document and keep it current. The change record must contain what requirements have been changed, why the change was needed, and what specifically was changed including the before and after wording of the change and the status. This includes if a requirement was deleted
- The change record must also contain who requested the change, who is responsible for making the change, and who will sign off or agree that the change is the proper change. Do not forget customer input in this process
- All changes must be tracked. The change log needs to record when the change request was made, initial estimated completion date, and the actual completion/change implementation date
- Each requirement must be a singular requirement in thought and be labeled in the document under a single requirement number or identification feature so changes can be documented and tracked. This also allows for design features or solutions to be tracked to a requirement
- When making changes, make sure that the new requirement is reviewed across the entire document to ensure that it does not conflict with existing requirements.

Good requirements are simple, verifiable, necessary, achievable, and can be traced back to a need. Requirements state what is needed, not how that need is to be achieved.

Requirements tell you the needed end results, not the method used to achieve those results. Simply stated, requirements meet the following four criteria:

- *Needs*: The requirement is necessary to meet a need of the program or the customer directly, and there is a consequence if the requirement is not met. Requirements must be traceable back to a need, request, or regulation. Requirements always state a need not a method or process on how to achieve a need.
- *Verification*: All requirements must have a quantifiably answer that can be verified and validated that it has been met.
- *Attainable*: All program requirements must be able to be met. All requirements must be technically and fiscally attainable. Requirements that are not technically achievable at the time of writing should be turned over to the research and development team so that the technology can be developed and implemented at a later time. If budget or schedule are issues, the team must decide to either shelf the requirement or adjust the budget and schedule to accommodate the potential solution(s). Don't write a requirement if you cannot afford it or if its gold plating the product. Give the customer solutions that they want, need, will pay for, and can afford.
- *Clarity*: Each requirement shall express a single objective that is clearly and simply written. A requirements statement must convey only one thought and not try to cover two or more needs in a single statement. Requirements must be written in proper format and text so that any potential confusion is avoided.

Requirements are a condition, need, or capability that a customer requires a solution to. Customers may be internal or external to the organization. Requirements provide the backbone of the program and will define program success. A well written and comprehensive requirements document that is developed at the start of a program will greatly enhance the chances of program success and help guide and maintain momentum during the program.

8 Developing and Understanding Strategy

Strategy is the plan that will drive the actions of an organization, a team, or an individual. A good strategy takes time to develop and requires research, reflection, homework, and the ability to ask tough questions in order to develop a plan that will act as a guide for moving forward. Strategy is the plan that guides the organization or team toward achieving specific goals and objectives even in the face of a constantly changing business environment.

The word strategy is simple enough—from a military (business) meaning it is: developing a plan that enables the effective meeting of the enemy on favorable grounds. Another simpler meaning is a plan that enables a desired outcome. Since the meaning of the word is so simple, why is developing and executing a strategic plan so difficult?

First, organizations fail to develop a clear vision that defines and gives meaning to the organization's strategy. A strategy must be driven by a vision that is clear, can be articulated, and is distinctive for that organization. The vision needs to answer four basic questions:

- How does the organization deliver value to its customers, employees, communities, and stakeholders?
- How is the organization differentiable from the competition?
- What gives the organization an advantage over the competition?
- Does it define the long-term objective of the organization?

Most leaders cannot clearly answer these questions distinctively, because the organization's vision is muddled. If the organization's vision is muddled, then the organizations operating plan and strategy will be disjointed. If the organizations leaders cannot state the vision clearly, mixed and confusing messages will be sent to employees, shareholders, and customers.

Developing a strategy takes time. Developing a strategy requires that time is made available to carefully think about the organization and the direction it is going and the direction it should go. Leaders need to take time to reflect, analyze data, think about the data, and then start formulating a plan that will turn into a strategy. A strategy defines the big picture for the organization. The following are a few questions that can help leaders define the big picture and the organization's strategy.

- Is the organization delivering to customers what we claim? Example: On time delivery every time?
- Is it clear how the organization delivers value? Example: Deliver fully certified and integrated structure on time with no defects.

- Do we understand clearly how we are differentiable from the competition? What are the five to ten capabilities that set the organization apart from the competition? Example: Composite and metallic fabrication, test, and certification capability.
- Would our customers agree with the answers to the first three questions, and are we funding and supporting the organization to reinforce the first three questions?
- Does the organization overall support the first three questions, or are we wasting time and resources in other areas and non-value-added work?
- What differentiates our competition?
- Can everybody in the organization articulate the organization's vision and strategy?

There are several ways organizations and leaders can answer these questions. First, the leaders need to step back, look at, and analyze the organization. The leaders need to take the time to step out of the daily running of the business and answer all seven questions, thoughtfully and completely.

Second, the leaders need to bring in others to double check their own answers. This can shed light onto issues, challenges, or opportunities that the leaders had not thought of in their own evaluation of the company and in their own answers. Leaders need to get out and survey the land and not totally rely on the input from others.

Many organizations do not value thoughtful thinking or reflection. Organizations have gotten so used to being reactionary and firefighting, that they no longer value the time it takes to do strategic thinking. Remember daydreaming is not strategic thinking. True strategic thinking is looking at data, performance reports, customer reports, and financial and cost data and analyzing that data to gain insights into what truly sets the organization apart and the condition of the health of the organization.

Individuals that are responsible for developing an organization's strategy need to step away from the constant firefighting and reactionary behavior that can distract them and keep them from moving the organization forward. In today's fast-paced business environment that is constantly changing, organizations will always need to respond to new threats and a changing landscape. To cope with this changing landscape, the organization needs to become flexible. Flexibility allows the organization to deploy resources to meet changing threats, while keeping an eye on the future. The organizational vision and strategy are the compass required to achieve success.

To help refine a strategy, leaders must bring others in, and allow open discussion and debate. Debate and discussion bring clarity to the strategy, holes or bad assumptions are brought to light and can be corrected for. Some leaders think that by constantly talking about something, it promotes clarity and buy in. In reality, people gain understanding and can internalize something when they can talk about and feel like they can own it. Getting others to own the strategy throughout the organization builds a stronger organization.

A major misconception many leaders and even individuals have is that strategy is only for the "top dogs" in the organization. Often first line managers, team leads, and even individuals do not believe that strategy is their concern. The truth is that strategic thinking can help anybody and is needed at every level of the organization. The company overall vision and strategy is for the senior leadership team to define, but teams and smaller groups within the organization need their own vision and strategy on how to support the overarching company vision. The key is that the team's vision and strategy must support the greater company vision and strategy.

Strategic thinking will set individuals apart from their peers and teams or departments above others. When an individual's or team's method of working clearly shows support for the bigger organization's vision and strategic plan, it sets them apart and in a better light, especially when budgets or promotions are being discussed.

STRATEGIC THINKING SKILLS/TACTICS

The following tactics can help individuals and teams think more strategically and build skills that support strategic thinking.

First get a heads-up mindset. Most of us have a heads-down mindset. We are so busy fighting the current fires and trying to complete that task that is right in front of us, that we don't take time to look up and see what's going on around us. In the wild, animals regularly stop and scout out the surroundings for opportunities and threats, their survival depends on knowing their surroundings, it's no different in business.

You must gain a solid understanding of your surroundings and the forces that are shaping it. What are the trends that drive the industry, what challenges, technologies, or forces are shaping the industry? To gain a solid understanding of the environment, individuals must:

- Examine your day-to-day work and look for trends, are there things that your peers mention every day? What causes them heartburn? What are the daily issues, challenges, or obstacles that you face that are constantly reappearing?
- Know how you and your team support the organization's bigger picture. Strive every day to ensure your efforts support that bigger picture, and when those activities don't, stop doing them. This requires time every day to stick your head up and look around
- Connect with others within and without your organization to gain a better understanding of the marketplace and to share ideas and insights. Conferences and professional societies are great places to share ideas, insights, and to gain valuable information on where the industry is headed
- Ask questions. Questions help you and others understand clearer. Questions help lay the foundation for a solid strategy. Ask yourself, where do I want to be in one year and five years? Does this specific task support the team's goals? The same questions can be asked by leaders to ensure organizational efforts support the organization's strategy
- Communicate clearly. People who can think and act strategically can communicate clearly. This is always harder than it sounds. People with good communication skills can explain clearly the issues, build understanding, and gain mutual support for the recommended solutions. To help achieve this goal, group your ideas. Make your points succinct. Prime the audience,

tell them up front why they are there and what you will cover, and then walk them through the process—issue/resolution-common understanding-options-agreement. Practice giving the main point and answer up front, and then walk the audience through the pitch back to the main point, resolution, and acceptance

- Make time to step back and think about the bigger picture and long-term plan. Long-term planning and vision are different than the daily short-range look discussed in bullet one. Bullet one is meant to help you get out of the firefighting mode, so you can take time to look long term. Take time to reflect and learn new things and look for new challenges
- Allow debate and conflict. This is a hard one, as we tend to look at a debate as a challenge and take it personally. Conflict can bring new ideas and insight. Focus on the issues and don't let the debate grow personal. Conflict is about questioning assumptions and the rationale that leads to those assumptions.

The skills discussed above help individuals and teams gain better strategic skills, and these skills transfer to the larger organization. But what are some of the activities that an organization (or an individual) can do to help hone a strategic plan?

- Look outside your organization for market trends and market insights. Talk to customers, suppliers, and partners. Examine their strategic plans and vision statements. Look at your customers' customers and see what is driving them.
- Conduct market research and competitive analysis to see what is driving your competition. Look at your competition's customers and supplier base. Look at your competitions research and development focus. By looking at the competition, you gain insight into the threat they pose, the direction they are heading, and also potential customers
- Look at the startups in your own and your customer's industries for insights and for potential shifts in the industry
- Attend conferences, trade shows, and events not only in your industry, but in other industries. This can lead to new business, greater industry understanding, and help identify trends or risks
- Look at proposals you have won and lost and talk to the customers and find out why you failed or why you stood out over the competition. This ability to conduct a post-proposal review internally and with the customer is extremely healthy for an organization. It helps the organization hone its skills to meet customer needs. The review may also reveal shortcomings in the organization or its products
- Question the status quo. Sacred cows hinder people from asking the hard questions. Questions lay the foundation that forms a good strategy and helps ensure that the strategy stays relevant in a fast-changing environment. A good rule of thumb has always been the five whys. A three-year-old asks why five times, so they get the basic cause of their original question answered. The five whys will lead you to the root of your first question or thought.

- Wisdom and innovation start by looking at the environment from different perspectives. Recognize the facts, and then reorganize the data that went into forming those facts to see if the facts become truths, and what factors are affecting those facts. When reviewing data, ask others to review your findings, this will help you solidify your understanding and will also help point out new avenues of thought or opportunities
- Once you have a strategy or vision developed, walk away from it for a while. Put some distance between you and the strategy, and let it simmer a little. Then after you have taken a walk or let it sit untouched for a few days, go back to it and see if you understand and believe in the strategy and if the assumptions you made are still valid
- Once a decision is made, stick with it unless you see data that points you in another direction. Make sure you can validate the data, and that it fits the organizational model and supports the organization's vision. Do not give teams or the organization whiplash by constantly changing the strategy or organizational vision. Leaders need to be flexible, but it does an organization no good if they are in a constant state of flux.

Strategic thinking is not just for the "Big Boss" or the heads of the company. Everybody can think about and benefit from thinking strategically about the organization and their own careers. All organizations need to have a vision and a strategy to achieve that vision. Strategic planning is not a black art, but it does take time and effort to pull together a solid strategy that can clearly articulate the vision and the path forward for the company.

9 Root Cause Analysis/ What to Do When Things Do Not Go as Planned

Problems happen, things break for no apparent reasons, and processes that were running well all of a sudden stop running. The following steps and discussion are good for solving most problems from a process issue, supply chain, or a production line shutting down. When life happens take a step back, take a deep breath, and then dig into the issue, but first remember these points.

- *Take ownership*: If you know there is an issue, raise it up and make it known to leadership. If it's a safety issue, make sure nobody can be hurt, and if necessary, get the line shut down. Safety should always come first. Don't jump to conclusions. You want to move fast to solve the issue and correct the problem, but don't jump to conclusions in a haphazard fashion. Gather the data you have, pull the team together, and move forward in a logical manner. Facts and data are the key to solving problems and ensuring that the right issue is solved and won't come back. The data may even point out other issues that need to be corrected, ensuring fewer issues in the future.
- *Gather the team*: Nobody has all of the answers. Pull a team together that can help you figure out what questions need to be asked, what data is needed, and the best path forward. Even if you're a "team of one" use your network to become your team, go to the experts and the people closest to the problem, and get their input. Teams solve problems and implement solutions.
- *Don't be afraid to ask questions*: You will never know everything, so ask questions! Questions help you uncover the truth and figure out what caused the problem and how to best solve it.
- *If in doubt, stop*: If in doubt, stop and ask questions, and go get the facts to clear up any doubt. Facts and data are the key to root cause analysis. The facts and data will drive the team and point you to the solution or solutions.

Do not try to sweep safety issues under the rug. In today's age of multimedia and built-in video capability, managing safety issues is of utmost importance. Do not assume you will be able to cover up issues that impact workers safety.

Root cause analysis (RCA) is a team-based logical approach to identify the root cause (the fundamental reason) for an issue and the best corrective action to take to ensure that the problem does not come back, while minimizing any further damage while you are working on finding the cause and implementing a solution. Companies may have their own name or process for root cause analysis, but the basics stay the same.

Most root cause analysis will follow a very similar path to the one given below. Adapt the following nine steps to meet your own needs. Recognize that your own organization may have their own format to follow, but it never hurts to look at several formats and processes and determine which will work best for you in any given situation.

1. Recognize that there is an issue/problem
2. Pull a team together
3. Define the problem
4. Define and implement a control plan (bound and contain the damage)
5. Determine the root cause by gathering and analyzing the data
6. Develop a correction plan, determine the path forward
7. Develop a prevention plan, ensure that once fixed, the issue does not come back
8. Implement and monitor the correction plan
9. Celebrate.

Now let's look at each of these steps in more detail:

1. *Recognize that there is an issue/problem*: This could be called recognize and admit that there is a problem. Recognizing and admitting are two different things. Recognizing the problem and bringing attention to the problem is the first step. You must get people to understand and admit that a problem or an incident has happened and needs to be investigated. Get leadership support to investigate the incident based on initial findings and data. At this point, you are establishing the need and getting buy in from others that a problem exists and needs attention.

2. *Pull a team together*: Problems such as engineering designs are solved by teams. Effective teams pull many viewpoints together and offer a wider range of experience than any single individual can have. Pull together a cross functional team, thus ensuring that you have all potential affected parties represented by the team. Your team may include people from operations, quality, design, test, shipping, marketing, and sales. If they are affected by the issue, they may have data that you will need. For instance, customer support may know that the pumps were failing in January last year right after the last major design change.

   A team does not need to be a formal team with 20 plus members. If you are asked to look into the problem and present solutions, then insure that you have the authority to ask all affected parties for input, data, and support as required. It is possible for a single individual to conduct a root cause analysis as long as they have the ability and authority to pull others in as needed.

3. *Define the problem*: Define the problem and, if possible, a timeline for the problem, the when, who, and what. When was the problem first noticed and who noticed it and clearly state what happened, the results of the event, and then develop a problem statement.

   For the last four shifts, 20% of cars have had scratches on the left rear door before they proceed into paint. The issue was raised by the second shift paint operator since it appeared again during his shift.

Defining the problem and the timeline helps the team narrow the scope of the investigation and also gives the team focus. Don't try to solve world hunger, make the problem something specific. If it's two issues, do two separate analyses, the issues may not be related, but brought up since people are now talking about problems.

When defining the problem, start out with why the team is being called together and what was the action or event that caused the alarm to be sounded (scratches on the door). Once the event is defined, start asking "why" questions to get down to the lowest possible definition and to start uncovering areas where more facts and data are needed. The problem statement should be simple and state the issue, who raised the issue, and if possible, how long the problem has been going on.

4. *Define and implement a control plan*: The control plan helps you stop any more damage from happening. The control plan does not solve the issue, it merely buys you time to find and fix the root cause of the issue. The control plan may be having all parts that are currently in process or completed inspected for the defect or shutting a line or machine down for maintenance. The control plan helps the team control or contain the damage and avoid bad hardware from finding its way to a customer. Always remember that a customer can be internal or external to the organization.

5. *Determine the root cause*: At this phase, you are gathering the facts and data and driving to a solution. Let the facts and data guide you to the solution, do not go into the analysis with a solution in mind. Keep an open mind and let the data speak for itself. Many implemented solutions have been the wrong solutions because the team let preconceived notions, ideas, or bias guide the process.

Data collection is a fact-finding mission, you're the detective. Talk to all stakeholders and people involved in the process. Look at all inputs and outputs from the operation. The issue may be traced back to incoming material and not the actual forming process. Listen to all inputs, and then try to piece the inputs into a logical timeline or story line. Let the data speak for itself, and do not try to bend it into a best fit story line. Above all, keep an open, inquisitive mind.

6. *Develop a correction plan*: Once the root cause of the issue/problem has been identified, then the team can develop a corrective action plan and fix the problem. A corrective action plan can include multiple elements, such as the speed and feed on the machine, the design of a holding fixture, as well as a change in the inspection plan to ensure complete part inspection.

7. *Develop a prevention plan*: Fixing a problem is one thing, ensuring it does not come back is something else. Gremlins that are not eradicated will come back. Always develop a prevention plan and communicate it to the larger team so that everybody can learn and prevent it from coming back. The prevention plan complements the corrective action plan. The prevention plan may include rewriting training plans or process planning. Prevention plans may also include a more robust inspection or maintenance plan. The correction plan fixes the problem. The prevention plan reinforces the correction

plan and makes the solution stronger by preventing it from coming back. The prevention plan is also called the defensive plan since it is the defense preventing the issue from coming back.

8. *Implement and monitor the correction plan*: After implementing the correction/prevention plans, monitor the process to ensure that the issue is truly resolved. Often, teams and management declare success and walk/run away. Monitor the situation and follow-up as necessary to ensure that the problem has been solved, and that the solution has not caused some unforeseen issue. At times, corrective actions uncover other issues. The old saying that "the water in the pond was lowered enough to expose more rocks (issues)" is very appropriate. There have also been cases where the corrective action put into place did not solve the problem, but just masked it or transferred it to another operation. Discussions with people on the line, inspection reports, schedule delinquency reports, and customer feedback are all great ways to monitor a system.

9. *Celebrate*: Take time out to celebrate with the team and to thank them for their efforts. Even if you're that "one-person team," take time out to thank the people that helped you find the data, determine the cause, and the corrective action, and those that put that corrective action in place. Remember to always give credit where credit is due. If you take all the credit yourself and don't recognize other efforts and contributions, you will not have the support when you need it next time.

RCA is a very powerful process that all technical professionals should understand. Problems will crop up. RCA is a proven way to uncover the cause of the issue and prevent it from coming back. It is a basic process and not hard in principle to follow, the challenge is to go deep enough that you truly get to the very basic or start of the problem, hence the name, root cause. Like a weed, you want to get the complete root out, so it does not come back.

10 Lean Engineering the Basics

When Lean manufacturing is discussed, it is often misunderstood. The first thing to understand about Lean manufacturing is that it is more than a tool or a process. Lean manufacturing is a way of thinking and acting. Lean is a way of living, and organizations that do Lean the best have made it part of that organization's DNA. At Toyota, Lean is part of the organization's DNA, the very core of the Toyota Production System. The Toyota Production System is based on a culture of continuous change and improvement. Toyota and other companies use Lean to drive out waste and improve customer value. To be truly effective, Lean needs to become part of the organization's culture, a way of thinking, not just a playbook or a set of formulas that can be brought out, dusted off, and used to fix a single problem, and then put back on the shelf.

Since I have stated that Lean is a way of thinking and is optimal when it becomes part of the organizational culture, does that mean that the Lean philosophy cannot be used on an individual cell or production line. Lean principles can be implemented in a single cell or production line and be effective. Individuals can implement Lean in their own work areas. By implementing in a single line or cell effectively, the value of Lean principles can be demonstrated and then expanded to the rest of the organization in a logical manner.

Converting an entire organization over has many potential pitfalls, especially when done too fast or with limited resources and experience. If Lean principles are implemented too fast or in too wide of a scope, the initiative will have limited results and face a lot of unneeded resistance. The best approach is to start small, show success in a single cell or product line, and then win others over and expand it once the concept has a foothold within the organization. Cultural change takes time and energy. Win support by showing success, and then win others over by training, open communication, and by showing firsthand the power and potential Lean practices have on the success of the organization.

THREE BASIC PRINCIPLES OF LEAN

The first principle of Lean is the Japanese word "Kaizen" or continuous improvement. One reason that some organizations are the best at Lean is that they have made Lean part of their culture. They understand that Lean is a journey with no end, that there is always a better way, and that the competition is not going to stop, so they cannot stop. There is always room for improvement and to learn.

The second principle of Lean is people. In an organization that has truly adapted Lean, everybody has a voice. The people that are closest to the problem will have the best ideas to solve the problem, if given the proper training and the ability to voice their ideas. The organization and its leaders must value their people and their capabilities and invest in them, so they can work in an environment that is based on continuous improvement (change).

The third principle of Lean is value from the eyes of the customer. Lean looks for ways to not only remove waste from the process, but to ensure that effort is used only on those activities that will add value to the customer, hence, those activities that the customer will pay for.

The three basic principles of Lean stated in a simpler form, Kaizen, People, and Value:

- *Kaizen*: Continuously improve yourself and the process. There is always room to improve. Lean is a journey
- *People*: People are the most important resource an organization has. Value and respect your team, the team will know where the issues are and will provide the solutions, if given the training, support, and permission
- *Value*: For a discussion on lean, value will be defined as what the customer will pay for. Every step in the process needs to add value from the viewpoint of the customer. If the step does not add value, then it is waste and needs to be removed.

Now that we have defined what Lean is and the three basic principles of Lean, what does it mean to an engineer or manager out on the shop floor or leading a manufacturing design team? How can Lean be implemented on the floor in a single cell or production line? Both of these are very good questions and questions that have simple straightforward answers.

First don't try to complicate it or over analyze it. Lean is not hard, it's based on the three basic principles mentioned above and has a toolbox full of tools that can be utilized to improve efficiency. Since Lean has many tools, individual tools can be chosen for specific utilization or situations. The team or the engineer can pick and choose which tools will work best for the specific situation.

Lean does not require a six-sigma black belt or a consultant to implement. Six-sigma black belt training is very powerful, but is not required in many situations since Lean principles are built around simplicity. Lean is meant to be simple and kept simple; six-sigma analysis methods may help the team quantify options and solutions, but are not needed to get started or to improve a process. The purpose of the following discussion is to point out tools that teams and individuals can use immediately to improve processes and help them get started on their own Lean journey. Employing a Lean practitioner or Lean Black Belt can be a great assistance to the team and can speed up the Lean journey, but is not required. The best approach to Lean at first is to keep it simple and, if necessary, add complexity and statistical analysis methods as required.

CONCEPTS OR PRACTICAL APPLICATION OF LEAN PRINCIPLES

Muda

A key concept of Lean is "muda" or the elimination of waste. Customers will not pay for waste; anything that does not add value to the product from the customer's perspective is waste and needs to be removed. There are eight traditional forms of waste in manufacturing systems, with a ninth form added below. The potential forms of waste are as follows:

- *Over production*: Producing more product than the customer wants or sooner than the customer needs.
- *Inefficient work in process*: Having too much product in work or raw material stored in a warehouse is a major source of waste. Inefficient work in process can have a very bad side effect, it can hide other inefficiencies in the system by providing a manufacturing buffer that hides issues, such as poor machine uptime. Inefficient work in process can hide bottle necks in the manufacturing system or material handling or ordering systems:
- *Waiting*: People, parts, or machines that are waiting for data, raw materials, stock, or other parts in order to begin or complete a cycle of work.
- *Motion*: The unnecessary movement of material, parts, or people to complete a task or process. This may include, bending, turning, and running to get parts, data, or tools. If the motion is not adding value to the part or process, it is wasted energy and needs to be removed.
- *Transportation*: The unnecessary movement of material or parts between processes. Do not confuse transportation with motion. Motion is within a process, transportation is between process or cells.
- *Rework*: This is sometimes also called out of sequence work. Do the job right the first time and in the proper sequence; do not push problems down the line. Quality should be built into the parts design and the process.
- *Over processing*: Adding extra features or processing steps that the customer will not pay extra for. What does the customer require, and what does the customer consider value added? Teams should not over process or design products.
- *Wasted material*: Are you manufacturing to near net shapes, how much excess material are you putting into a scrap bin? What consumables go into your process, and how can you minimize those. Consumables include such items as bagging material, plastic bags, cardboard, masking material, anything that you use in the production cycle that is not part of the final product. Customers will only pay for the end product, not what is put into the trash or the scrap bin. This can also include packaging material, the material that you use to ship the part or prepare it for sale. Your customer is only interested in the final product, not in the shipping container.

These are the eight most common types of waste that are considered when evaluating a process. Lean is about people the ninth type of waste in most manufacturing systems that is often overlooked is:

- *Untapped potential*: Are you truly using your people and their talents optimally. Are your people engaged, using all of their talents, being listened to, and asked for input? Are you asking your teams and employees to bring their brains as well themselves to work? Are your teams being creative and innovative? Many companies say, "their employees are their greatest asset," but few truly make their employees believe it. Is there untapped potential in your teams?

APPLICATION OF LEAN PRINCIPLES

VALUE STREAM MAPPING

We now know the basic principles of Lean and the types of waste found in a manufacturing system, so let's now talk about implementation. The first step in Lean is to look at your process and do a value map of the process. A value map shows you in graphical form the processing steps that a part goes through before the customer sees it. The value map will tell you the steps that are adding value and the potential sources of waste that are within the current process. A value stream map is a tool to visually show product or process flow; it is the visual representation of the production cycle. A good value stream map will show areas of potential waste.

A value stream map will show several critical data points for each processing step. Three features make up the visual cues on the map, process boxes, inventory triangles, and process data boxes. The inventory triangle shows the amount of product in the queue waiting to be processed. The process data box shows key information about that process, such as set up time, speed of the process, if its automated or manual, and first time yield or quality and the process box. The process box represents a point in the process where inventory or hardware stops and can either have work performed on it or can be held in the queue. The process box describes the results or intention of the operation and the number of employees involved.

A simple value map is shown in Figure 10.1.

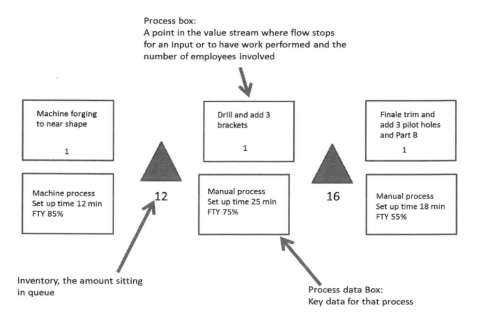

FIGURE 10.1   Lean manufacturing value map.

The value stream map will help you and your team understand the process and where potential waste and bottlenecks are within the process. Before you start making the value stream map, you should walk the process on the shop floor, so you fully understand the process and can account for all processing steps. After you walk the process, then you can layout the process and make the value stream map.

After you have the map made, re-walk the process with the map in hand to ensure you did not miss any critical information or steps. Also share the process with the owners on the shop floor, they can point things out that you may not know. For instance, when a tool is shared between multiple cells or how often they are held up waiting for material or a work order, all of these are steps in the process that may not be visible to the casual observer.

Once a value stream map has been established and the team agrees that all steps in the process have been defined, the next step is to look for waste in the process and remove that waste. Do not try to implement Lean before the value stream map has been established. As with any map, the value stream map helps ensure that you do not become lost, confused, or disorientated within the manufacturing or design process. Without the value stream map, you may implement Lean and be improving an area of the process that does not add value. Once the value stream map has been defined, then it is time to open the Lean toolbox and get to work eliminating waste and improving the process.

LEAN IN PRACTICE/THE LEAN TOOL BOX

There are over 20 Lean tools talked about in the literature today and in the "Lean textbooks." Let's just focus on seven of these tools that you can use in your own cell or product line today. Remember, many of these tools can be used in an office or design area or project just as well as out on a factory floor.

The seven tools we will look at are not placed in any particular order. The tools listed can be utilized on any product line, so you can pick and choose which ones to use to get your team moving on their Lean journey. The seven tools discussed below are 5S, Andon, Gemba, Kaizen, Poka-Yoke, Standard Work, and Total Productive Maintenance (TPM).

5S

The simplest and one of the most powerful tools in the toolbox is 5S. 5S is a logical way to clean up the area, organize the work area, and keep it clean. A clean well-organized work area will assist in product flow and help identify other areas of waste. 5S goes with the old saying "a place for everything and everything in its place." The five S's are:

- *Sort*: Eliminate what is not needed in the work area to complete the specific task of that work area
- *Set in order*: Organize all remaining items so you can see all tools and parts easily and identify them
- *Shine/sweep*: Clean the area and keep the area clean. Also inspect the area for safety hazards or potential issues with the machines, tools, and equipment. Do this step after you have done the first two and get rid of anything not needed

- *Standardize*: Write standards and processes for all work done in the area, including the following, standard maintenance in the cell, such as cleaning and safety walks. Make a clean, organized work area a standard
- *Sustain*: Apply the standards on a daily basis, and enforce the standards until they become part of the normal culture and thought process.

5S will help the team eliminate waste caused by a poorly organized work area, such as wasting time looking for tools, equipment, or hardware. A clean and organized work area is also a more enjoyable area to work in.

ANDON

Andon is also called the visual cell or factory. There are two parts to Andon. First is the visual part, which is no more than a simple way to show the condition of the cell. In its simplest form, Andon can be a set of colored lights or flags, green when things are going as planned, yellow when the cell is waiting for something, and red when there is a problem or work stoppage. A key element of Andon is that any employee has the power to change a light from green to yellow or red without asking for permission. The lights are intended to trigger a rapid response, so that the cell can get back up and running; this can be calling a manager or crew chief directly.

Andon should include a second element and include a team board, which shows the status of all activities that the team has in work from projects, team members on vacation, scheduled training, scheduled maintenance, and work schedule including units to ship. The combination of the lights and a status board coupled with team empowerment can be a very powerful tool ensuring production is flowing and quality is being maintained. For Andon to be effective, the team must be empowered for the productivity and ownership of the cell, they should be given and held responsible for the productivity, quality, and efficiency of the cell. Remember Lean is about being simple, so keep the status boards simple and visible.

GEMBA

There is an old saying on factory floors "real work doesn't get done in a manager's office." Gemba is a reminder to get out on the factory floor and see what is happening. Gemba promotes interaction between team members on the factory floor that includes engineers, managers, and the staff out on the floor. The best way to understand a process is to see it firsthand.

Gemba can be reinforced by having a 15-minute meeting at the start of every shift at the cell. This meeting should include the crews of both shifts, as well as all support staff. In this meeting, all issues can be brought up, maintenance concerns, quality schedule, and cell status. This meeting should be run by the cell members. Leadership is there to understand status and provide support when needed.

Get out of the office and see things firsthand, and don't rely on information from others when you can get it firsthand. Don't be afraid of interacting with the shop personnel. People love to talk about their jobs, what they are doing, and what bothers them or is giving them problems. Knowledge is critical for Lean to work and firsthand knowledge can never be replaced.

KAIZEN

Kaizen is the heart of Lean. Kaizen means continuous improvement. Great improvements and strides in efficiency could be achieved by taking many little steps and always looking for ways to improve. Kaizen is not an individual effort. Effective Lean programs involve everybody and ensuring that everybody has a voice. To continuously improve a process or a factory, listen to every person on the team and look at the problem or process from their viewpoint. The combined talents and efforts of the team will be required to build a Lean culture.

By implementing Kaizen and actively living the philosophy of continuous improvement, a Lean culture can be built. Remember, for Lean to be effective, it must become a way of life. You can never stop improving, for when you do, your competition will overtake you. A few simple ways to start building and instilling the Kaizen way of life within your organization are:

- *Have a suggestion box, use it, and give simple rewards and recognize suggestions.* Your employees know what's wrong and will give suggestions, if they think they will be listened to and respected for their efforts and opinions. Many suggestion programs fail because the organization does not provide feedback and does not budget money or resources for improvement programs. If employees do not think their suggestions are going to be taken seriously, then they will not even try to think of ways to improve or bring up the idea when they do have it. The worse thing an organization can have is employees that are just keeping their heads down doing their jobs because they believe management and the organization does not care.
- *Have Kaizen workshops in the cell.* These events don't need to be week-long events, but short half-day, one-, or two-day events built around ideas generated by the individuals in the cell or by opportunities/challenges from others outside of the cell. For example, a recurring quality issue, a safety concern, long periods of downtime all can be triggers for a Kaizen event.
- *Look at your cell or your factory/office area and fix what bugs you.* Take the simple, easy things first, but take a hard look at the process or environment and fix what bugs you. Always look for the simple solution and remember if the initial idea does not hit the mark 100%, then change and adapt. Kaizen is continuously taking small steps to improve performance and efficiency.
- *Get started, don't just wait for permission.* The elimination of waste and lost talent, creativity, or morale is an issue that Kaizen can help with. By involving everybody and telling (and showing) them that they are being listened to and respected and the intent is to build a better workplace and organization, the idea of continuously improving can take hold.

POKA-YOKE

Poka-Yoke is error proofing. The idea is to error proof, also called "Murphy Proof," the design or the process. Poka-Yoke is a way to drive to zero defects. Inspection of a part only finds mistakes after those mistakes or errors have happened, rework is expensive and ties up resources that could be better utilized making good parts or working on the next product. Rework is wasteful and goes against Lean. Poka-Yoke looks at both the design and process to analyze where the defect is happening, the root cause of the defect, and finding a solution to eliminate the defect.

Elimination of the defect may require a design change, such as making a round hole slotted so that the part can be loaded into the jig one way. Elimination of the defect may require adjusting the speed of the machine or the way parts are loaded to reduce the risk of jamming and damage. The key to developing and achieving a Poka-Yoke process is a simple and logical approach:

- Define what the defect is, scratches, misdrilled holes, incorrect forming of the part, or incorrect placement of a label, for example
- Look at the part and the process and see where in the process the defect is occurring
- Determine if the defect is caused by machine, human, or human and machine interaction
- List all potential causes of the defect, and examine each potential cause separately
- Look for multiple ways that the cause of the defect can be eliminated
- Determine the best approach(es) forward. In some cases, more than one solution will be required to make a robust error proof process. For instance, it may require turning one hole in the part into a slot, so it can be fed into the machine only one way, and it may require that not only is the tool redesigned for the slotted hole, but that the tool be anodized to prevent galling, which is also causing defects
- Keep an open mind and listen to all voices
- Once a corrective action has been taken, re-visit to ensure that the errors have been eliminated, and that the corrective action has had the intended results
- Share the story with others. This shows respect to the team and appreciation for the work they did. Sharing the story of the journey shares the knowledge and lessons learned with others, thus helping to instill Lean in the culture and enhance cross-functional, cross-team knowledge.

Poke-Yoke is a great way to reduce waste and improve overall quality.

STANDARD WORK

Standard work is a Lean tool that is often overlooked. One concept of standard work is to just allow one type of work or part through a cell. Limiting a cell to one type of work or part may not be practical in all shops and may also not utilize shop resources to their full potential. Also limiting work traveling through a cell is an oversimplification of standard work.

Standard work is the reviewing of work practices and procedures and defining and then documenting the best practice for a specific operation or job. Once a procedure has been defined and documented as a best practice, it should be disseminated throughout the shop. The best practice is meant to act as the standard work instruction for a specific task or job, ensuring that every employee does the task the same way every time, this reduces variation in the production or design process.

Once defined, standard work instructions are not static documents, as newer practices or procedures are learned or refined, the document should be updated. A best practice or standard work instruction document is a living document. Standard work instructions are the baseline and the common reference point for current production and future improvements to the process.

TOTAL PRODUCTIVE MAINTENANCE

TPM, also called preventive maintenance, is often times overlooked, and by many not considered a Lean tool. When preventive maintenance is not done regularly, machines break down or become unreliable. TPM is a complete approach to maintenance that focuses on proactive preventive maintenance that in the long run enhances machine availability. By setting all machines up on a routine maintenance schedule, maintenance can be carried out in a logical manner. Maintenance procedures are often times broken down into small operations that could be carried out at different times and spaced such that production is not hindered and the machine is kept in top performance.

One focus of TPM is that machine operators are trained and treated as owners of the machine. Teach the operators to perform some routine maintenance procedures. By training the operators to perform routine cleaning and maintenance procedures, the operators are instilled with ownership of the machine and the process. A good TPM program can greatly increase shop efficiency by increasing uptime, reduce cycle times, and eliminate machine downtime and defects caused by poorly running equipment. All shops should have all of their equipment on a TPM program and involve the machine operators in the daily upkeep of their equipment. Ownership is a key element of effectively implementing Lean and creating a culture that embraces Lean as a way of life.

LEAN IN SUMMARY

The tools discussed above are just a few of the tools in the Lean toolbox, but they are tools that are easily understood and can be implemented in any cell, shop, or work environment and do not require extensive training to understand or implement. Lean is a participation sport and requires involvement, so don't wait for formal training, decide to start your own Lean journey, and follow the simple guidelines above.

Lean is about the process, removing waste, and the most important part of any shop, its people. Lean principles enable and support the shop personnel. People are at the heart of Lean, listen to the team, and let everybody have their input and amazing things will happen. Lean is built on trust and open communication. People need to feel free to talk and express themselves and believe that they are valued for Lean to work. Remember your best ideas for Lean will come from the floor and interaction on the floor. If the shop personnel know you trust them and that they are valued, they will respond in a positive manner, and Lean will take root.

Lean is a journey, a never-ending journey. The elimination of waste and adding value to the process that the customer is willing to pay for is never ending. The goal of Lean is zero waste and optimal value for the customer. Lean is a logical process to remove waste, improve quality, and remove variation. Open communication and teamwork are critical for Lean to work. Lean is about identifying and removing waste within the system. Don't let all of the hype about six sigma scare you, look at the basics and pick one or two of the simple tools and start moving, as you get more familiar with Lean principles, then you can add more complexity. The basis of Lean is simplicity and logic. When you deliver optimal value to the customer, you will also deliver optimal value to the organization.

11 Cost Estimation/How Do You Figure Out the Cost of Manufacturing

Should-cost means what should a part cost to build, not what you charge or get charged. It's used in contract negotiations and estimates.

Engineers, managers, product owners, and sales teams need to be able to analyze what it costs to make a product or gadget, this cost estimation is often referred to as a should-cost. Programs use cost estimations for two purposes. The first is to give a good idea of what a part should actually cost to make using a specific manufacturing process. The second is to look at areas where teams might be able to get cost out of a part. A good should-cost estimate can help the team understand where the hidden costs of a part lay, and if they are being potentially overcharged or bidding incorrectly on a project. Hidden costs of a part may be in material, the manufacturing process, direct labor, or indirect labor.

If done properly, a thorough should-cost estimate will help the manufacturing team understand what a part costs to make and how those costs could be broken down. A should-cost estimate could also give a team insight from both a time and cost perspective into what each operation within the build process takes; this time/cost analysis could give the team true insight into the cost/manufacturing process for a part. A good should-cost estimate is a bottoms up review of what it takes to make a part or an assembly.

The steps to conducting a good should-cost estimate are straightforward:

- Pull the drawing or the model and understand the part from the design engineer's perspective
- Understand the requirements shown on the drawing. Read the drawing notes and associated specifications or special process notes
- Read the process planning, build plans, and inspection plans
- If the part has been made, talk to the technicians and shop floor personnel who made the part, ask them the following questions:
  - What causes them issues in the build process?
  - Areas of concern with the design or build process or planning?
  - Do a time study or ask the technicians how long each processing step takes
  - Did they use any special tools, tricks, or shop aids that are not called out in the planning or process sheets?
  - Always ask if they have suggestions about how to make the part better or make the fabrication process more efficient.

The following list can be considered when gathering data for the should-cost estimate. This list should only be used as a starting point and not as a complete unmodifiable list. Modify the list to fit specific processes, parts, and requirements. The list is intended as a starting point; a template of sorts to begin the cost estimation. Composite materials and processes are used in the example below, the same approach can be modified for different materials and manufacturing processes, additive, plastics, or metallics.

- *Material used:*
  - What is the type of material: thermoplastic, thermoset, or a ceramic?
  - Type of fiber used, Kevlar, carbon, fiberglass, ceramic, or a hybrid (glass/carbon)
  - Type of fiber used, 1k, 2k, 12k yarn, plain weave, 8 harness satin, braid, or 3d preform
  - Type of secondary adhesives used in the fabrication of the part and how much
  - Is core or foam used in the fabrication of the part (and type)?
  - Are specialty materials used, such as erosion films, seals, or wear strips?
  - Are inserts or fillers of any kind used in the fabrication of the part, and, if so, their cost both in material and fabrication?
  - Paint
  - Disposable materials, such as breather films, bagging materials, thermal couples, and release films.

- *Manufacturing process:*
  - Hand layup
  - Resin transfer molding
  - Compression molding
  - Pultrusion
  - Automation, fiber placement, filament winding, tape placement.

- *Part geometry and complexity:*
  - Curvature, single curvature, double curvature, or compound curvature
  - Features, such as holes, flanges, bevels, reinforced areas, webs, or stiffening members
  - The number and placement of ply drops
  - Number of ply's, size of the ply's, and amount of material used
  - The number of different materials used in the part.

- *Processing of the part:*
  When evaluating the processing (manufacturing) of the part, account for all processing steps including secondary processing steps, such as bonding or paint. Different processes will require different amounts of touch labor and potentially different burden or overhead rates for different shops or operations. When evaluating the processing steps, look at the details in each step and the number of steps to complete the entire process:

- Ply kitting
- Ply placement (hand, automation, combination)
- Bagging or preparation for cure
- Autoclave
- Presses
- Ovens
- Preparation for bonding or painting
- Secondary bonding
- Painting
- Inspection
- Tool preparation (this may include several tools).

- *Special processes:*
  Do not leave out any special processes that might be required to fabricate the part when conducting the cost review. The way to identify a special process is to look at the drawing and the specifications listed or what is called out in the notes. Discuss with the design engineer any special process that have been called out on the drawing to ensure that the design engineer understands what he is calling out and to ensure that the part being defined by the drawing is truly what is being produced and what the design engineer wanted in the first place. Missed specifications or call outs have caused major issues including quality escapes, parts not meeting design life requirements, part failures, and increased manufacturing costs. Use the cost review as an opportunity to review the drawing and the manufacturing processes. A detailed review could lead to better understanding of the part and cost out or cost avoidance opportunities before production starts.

    The following are some of the special processes used in composite fabrication, metallics and other material systems will use some of the same special processes:
  - Painting
  - Bonding
  - Post cure processes
  - Secondary heat treatments, burn out, or stress relieving cycles
  - Special or non-standard cure cycles
  - Grit blasting or special secondary preparation operations
  - Application of special or specialized materials during the lay-up (peel ply, erosion films, or wear strip)
  - Disposal or recycling of the excess material. Some materials have special disposal requirements. These costs are generally found in the overhead cost, but not always
  - Machining or trimming of the parts
  - Special coatings. Some parts may require special heat or thermal coatings
  - Inspection and quality processes or inspection, NDI, coordinate measuring machines reports
  - Special quality or testing requirements, such as testing witness panels used for mechanical testing, mechanical or destructive testing used for part verification or quality assurance, or material sampling

- Logistics, the cost of shipping the part for secondary operations or delivery to the customer
- Planning and shop engineering support.

Be sure to step through each process and record the time it takes to make the part from step one until it is delivered to the customer. Ensure that all excess, trimmed, or scrap material is accounted for. Account for material utilized and every minute of manufacturing and touch labor time even in the secondary operations, such as tool preparation.

- *Shop overhead rate:*

The overhead rate is the rate you apply to each manufacturing hour or touch hour on the part. Any time a technician touches the part, there is a cost applied to the part. Overhead rates usually include the standard salary and benefits of the technicians, utilities, building costs, and support function costs for the site.

Consider the shop rate the cost to turn the lights on in the shop so a part can be made. The shop rate will generally also include shared equipment, such as autoclaves, ovens, coordinate measuring machines, or large machining centers. Shared equipment is used in the fabrication of multiple parts or processes and the cost of the equipment that is shared by multiple programs.

Shop rates will vary from site to site and country to country, so make sure you know the dollar value for the specific site that the part is being evaluated against. In most cases, this number can be provided by finance, plant leader, or manufacturing engineer. In some shops, the rate will vary from one area of the shop to another, military, small part fabrication, machine shop, or inspection. If the shop rate varies, apply the proper shop rate for each specific process being evaluated.

- *Tooling cost:*

All good should-cost estimates must include tooling costs. The cost of the tooling and tooling aids including machine programs are often overlooked in a cost estimation. The cost of the tooling should be amortized over the number of parts being made. If the tools are being bought under a contract, this should be stated and the cost not applied to the cost of the specific parts. If the customer is paying for all tooling upfront, leave it out of the should-cost estimate, but still capture the costs. This information can be used for program cost estimation efforts on later programs or during program audits.

Always make sure you have a complete tool list to ensure all tooling costs are captured. The tooling list must include all shop, tooling aids, disposable tools, such as scissors, resin cups, and mixers. When looking at the tooling cost, consider the following for each tool, tooling aid, fixture, or program used in the fabrication process:

- Hours to design the tool if not included in the tool fabrication cost
- The cost to have the tool fabricated

- Tool inspection and or validation (always inspect the tool before it is placed into production to insure it meets design requirements)
- Account for each tool required to manufacture and sell the part (this may not be a complete list):
  - Material cutting templates and programs (electronic or physical)
  - Laser templates
  - Layup tool or mold
  - Part specific carts or handling fixtures
  - Inspection fixtures
  - Inspection programs
  - Machining programs
  - Machining fixtures
  - Tooling aids, secondary tools used in the fabrication process.

A good should-cost estimate is a tool that provides the program and manufacturing teams understanding of the cost drivers for a part or a production process. The should-cost estimate can be used to help drive cost out of the part or as a baseline to negotiate the fabrication of the part by an outside vendor. The should-cost estimate data can also help engineering understand what is driving the cost of the hardware, so design decisions can be based on not only structural or performance requirements, but also manufacturing requirements and cost.

A good should-cost estimate is a very powerful tool for manufacturing and programs teams to have, especially for complex or high value parts. Once completed, should-cost estimates could be used in the following fashion:

- Part price negotiation: both as a supplier and a buyer of the hardware. A good should-cost can help determine what the value (cost) of a part is both from a seller and a buyer viewpoint
- A complete should-cost evaluation can be used as a starting point for a cost savings initiative
- Used as a tool to help the engineering team understand cost drivers in the shop and with different manufacturing processes and materials
- A set of several different validated should-cost estimates can be used to establish a set of standards for cost estimation in advance design or product improvement teams
- A part family should-cost or a good should-cost model could be used to help the team benchmark the competition.
- A good should-cost could also be used to determine if a re-design of a part can truly make it cost competitive or if a better option is to bring out a new product or new design.

Manufacturing teams need to know what goes into building the hardware they are responsible for. Once an engineer or a team understands what goes into the hardware, they need to know and understand the cost drivers that affect that hardware so cost saving ideas can be developed and flushed out. Should-costs estimates are a tool that many teams and engineers overlook. The utilization of should-costs estimates can help drive a culture of continuous improvement and cost savings. Should-cost estimates help teams determine what parts need cost out initiatives and what processes are cost effective. Good should-cost estimates help teams make data-driven decisions, from a buying, selling, and cost or process improvement perspective.

Closing Thoughts

In today's world, everything is changing and in flux. Technology is moving at an ever-increasing pace, new materials and processes are always being invented and tested out. We must always remember five key principles as we navigate this constantly changing landscape:

- Technology may advance, but the basics stay. This handbook is intended to help give you the basics. The basics of sound engineering and logic coupled with hard work will get you out of a lot of tight spots
- Never stop learning. The world moves ever faster, and as it does it opens opportunities to explore and learn. Knowledge and the willingness to learn and try new things are very powerful
- People, and how you work with them, will be a factor in your success. Respect and understanding of others go a long way
- Always remember that combining different views can lead to the next breakthrough. Listen to others and observe. You learn more by listening and asking for clarification, than you do by talking. Inspiration and understanding can come in many forms
- Give credit where credit is due, recognize the success of others, and appreciate the help you receive along the way. Relationships matter as much as technical expertise.

I hope this handbook serves you well on your journey.

Good luck in all of your endeavors.

Kevin Retz

Index

Note: Page number in bold refers to table.

A

action log, 5
active listening, 58–60
Andon, 84

C

commitment, 2, 8–9
cost estimation, 89
creativity, 43, 45
critical thinking, 53–55, **56–57**
criticize, 4

D

difficult conversations, 14–16

E

email, 33–34

F

feedback, 11–14, 18, 55, 85

G

Gemba, 84
goals, 21, 63, 65

I

innovation, 43
integrity, 4, 7

K

Kaizen, 79, 85

L

Lean, 79, 87–88
learning, 16–17, 47, 58

M

mentoring, 47
Muda, 80–81
multimedia, 41

N

Newton's third law, 2

O

objectives, 63–64
overhead rate, 92

P

passion, 3, 18–19
Poka-Yoke, 85–86
praise, 4
presentation, 37, 40
purpose, 10, 18–19, 21–22, 29

R

recognition, 49
reflection, 8, 10–11, 13, 16
relax, 40–41
represent, 8
requirements, 63–68
respect, 1, 4, 6, 9, 16, 24
Root Cause Analysis (RCA), 75

S

schedule, 2
should-cost, 89
special process, 91–92
standard work, 86–87
strategic thinking, 71–73
strategy, 48–49, 69
stress, 9–10, 18

T

team website, 31–32
tooling cost, 92–94
Total Productive Maintenance (TPM), 87
total team, 23–27

V

validation, 66–67
value stream mapping, 82–83
virtual teams, 29

Printed and bound by CPI Group (UK) Ltd, Croydon, CR0 4YY

23/10/2024

01778243-0009